Studies on Contemporary Europe

Edited by
PROFESSOR ALAN S. MILWARD
University of Manchester Institute of Science and Technology

I
Budgetary Politics: The Finances of
the European Communities

Studies on Contemporary Europe

Budgetary Politics: The Finances of the European Communities

HELEN WALLACE
Civil Service College

University Association for Contemporary European Studies
George Allen & Unwin

First published in 1980

GEORGE ALLEN & UNWIN LTD
40 Museum Street, London WC1A 1LU

Satellite picture of Europe reproduced courtesy of the Commission of the European Communities

British Library Cataloguing in Publication Data

Wallace, Helen
    Budgetary politics. − (Studies on contemporary
    Europe; 1).
    1. Finance, Public − European Economic
    Community countries
    2. European communities
    I. Title II. Series
    336′.09168   HJ1000

    ISBN 0−04−382023−9
    ISBN 0−04−382024−7 Pbk

Typeset in 11 on 12 point Plantin by Computacomp Ltd,
Fort William, Scotland
and printed in Great Britain by A. Wheaton & Co., Exeter

Contents

List of Tables

Editor's Preface

The University Association for Contemporary European Studies (UACES) exists to promote the study of contemporary European society in all its aspects. To do so it brings together a large number of scholars from many different disciplines. One motivating force for these scholars has been an awareness of the inadequacy of their particular scholarly discipline to provide satisfactory answers to the complex problems which they were handling both in research and in teaching. Many aspects of contemporary European economy, society and politics are indeed hard to illuminate if the light comes from only one of the traditional disciplines of academic study and this has meant that teachers, students and everybody else are frequently without adequate information on topics of immediate and important interest. The Association has therefore commissioned scholars currently working in such areas to present in short form studies of problems which are of special importance, or specially noteworthy because of the lack of easily accessible information about them in current public and academic discussion. The studies are written by experts in each particular topic. They are not, however, merely for teachers and students, but for anyone who may wish to find out something further about subjects which are now much discussed but about which real information is still hard to come by. In this way the Association hopes it may bring closer what are often the separate worlds of academic and public knowledge while at the same time providing a service to readers and students in a relatively new field of study.

ALAN S. MILWARD
*University of Manchester Institute
of Science and Technology*

Author's Preface

This study was made possible only because numerous officials and academics were willing to provide information on and interpretations of the financial activities of the European Communities. Particular thanks are due to them all for their help and forbearance, though needless to say the responsibility for the views expressed and for the accuracy of the detail rests entirely with the author. Thanks are also due to Marie Walsingham and her colleagues for the care and patience with which they typed and retyped the manuscript. Last but not least a special debt is recorded to Harriet Wallace, William Wallace and Ann Pendergrast for the remarkable good humour with which they tolerated the stresses caused by the production of this study.

HELEN WALLACE
Civil Service College

List of Abbreviations

CAP	Common Agricultural Policy
DG	Directorate General of the Commission
EAGGF	European Agricultural Guidance and Guarantee Fund
EC	European Communities
ECSC	European Coal and Steel Community
ECU	European Currency Unit
EDF	European Development Fund
EEC	European Economic Community
EIB	European Investment Bank
EMS	European Monetary System
EMU	Economic and Monetary Union
ESF	European Social Fund
EUA	European Unit of Account
Euratom	European Atomic Energy Community
GDP	Gross Domestic Product
MCA	Monetary Compensatory Amount
OECD	Organisation for Economic Co-operation and Development
RDF	Regional Development Fund
UA	Unit of Account
VAT	Value Added Tax

I Problems of Analysis

'Who gets how much' and 'who pays?' are always crucial questions in any political system. Arguments about financial resources, the purposes to which they are put, the pattern of their distribution, and their susceptibility to democratic control are regularly among the main preoccupations of local and national politics. Indeed many of the standard works on political science regard the controversies over the allocation and redistribution of resources and their resolution as identifying characteristics of politics in the broadest sense.[1] The budgetary process offers a prism through which to distinguish underlying values − political, economic and social − in so far as they are reflected in specific decisions about the raising of revenue and the distribution of expenditure. Cash values reflect political values. New governments on taking office expect to distinguish themselves from their predecessors in their handling of public finance. One political system is marked out from another at least in part by the financial weight accorded to different sorts of expenditure, and indeed by the balance struck between public and private wealth. As the size of the public sector has grown in countries with mixed economies, so these issues have acquired increased political salience.

Equally for economists budgetary questions have major significance. Budgets provide instruments through which the economy is managed, with measures designed to allocate, redistribute and stabilise economic resources.[2] The larger the public sector, the greater the impact of budgetary decisions on the economy as a whole. As the parallels between these political and economic concerns become more evident, so the issues that arise from public financing have become of prime interest to the growing number of students of political economy.

Studies of policy analysis have also begun to generate considerable interest in the budgetary process. Obviously the mechanisms adopted in any system of government to handle public finance shed considerable light on the functioning of that system, on its institutional relationships and its capacities. Thus the student of administrative behaviour has much to learn from the analysis of public finance, particularly by examining the influence of administrative factors on the choice and the effectiveness of expenditure programmes.[3] There is now a substantial literature, largely American, on the budgetary process itself.[4] Danziger's volume is a useful contribution, in that it attempts to test a variety of theoretical approaches to budgeting through a series of case studies based on local government in the UK. In particular he draws on the models of rational policy-making, incrementalism and organisational analysis and then explores the utility of econometric models and of an environmental or 'demographic' approach. For anyone interested in pursuing theories of budgeting in their own right the book offers a number of provocative insights that could with profit be drawn on in further research on the budgetary process in the European Communities.

More immediately relevant, however, for an enhanced understanding of budgetary processes is the volume by Wildavsky, since its primary purpose is to develop a comparative theory. It aggregates research findings on the development of budgets in several countries, both developed and underdeveloped, and on the use of new techniques in budgeting. Its empirical material includes studies of the management of finance within unitary systems (France and the UK) and among different levels of government (primarily in the USA). Wildavsky recognises at the outset that 'budgets are attempts to allocate financial resources through political processes to serve different human purposes'. Thus inevitably they emerge from conflict and contain a composite of features designed to serve a variety of needs. Equally important, 'a budget is a record of the past', in that it encapsulates the outcomes of battles already fought; it is also 'a statement about the future', in so far as expenditure programmes may attempt to alter behaviour in the future. Wildavsky's own position is that of an incrementalist. Once a budget has been established it provides a base from which change is generally made through an incremental process, whereby this year's budget is related to last year's with only a narrow range of increases and decreases. The base budget builds up a pattern of expectations and behaviour by both the providers and the recipients of revenue, whereas new programmes operate in an environment of uncertainty that can introduce further strains and tensions into the process of allocating resources. It is important also to recognise the different roles of what Wildavsky terms the 'advocates' of expenditure (the claimants) and the 'guardians' (the controllers) of the public purse. Though these features are common to most budgetary processes, there are variables that differentiate them from each other; most important for Wildavsky, these are the relative wealth of the system and the degree of predictability over outcomes. These variables thus permit the identification of different types of budgetary processes: from the more narrowly based, associated with poor countries or units of government that have uncertain political rules, to the more sophisticated, associated with wealthy countries and a more stable decision-making process.

International Comparisons

A key assumption that underlies most discussions of budgets and public finance is that their scope and effectiveness depend on their location in tightly knit political and economic systems. A certain degree of centralisation of decision-making and commonality of values is necessary to permit the pursuit of political and economic objectives through the use of budgetary instruments. Obviously at a national level there are important distinctions to be drawn between unitary and federal states, in that the former in principle are more cohesive and thus in principle allow highly structured budgetary processes. Conversely, in a federal system or one where there is a significant degree of decentralisation, the management of public finance is more dispersed and fragmented. It is self-evident that the international system is very underdeveloped and unsophisticated by comparison with both. By and large international economics and politics have not been bound up with the determination of choices about the allocation of financial resources, let alone with redistribution. There have been mechanisms designed to afford crude measures of stabilisation and a limited degree of resource allocation and redistribution. But the absence of an entrenched framework through which such choices might be expressed has meant that these issues have obtruded only at the margins or in times of international crisis. Examples of this include the Marshall Plan and some of the measures adopted internationally at the time of the 1973 oil crisis. Consequently, claims, for instance, from developing countries for a New International Economic Order are unlikely to achieve any permanent change without the creation of internationally accepted redistributive mechanisms. Similarly, attempts to meet commodity shortages on a global scale seem doomed to remain *ad hoc*, if made without the underpinning of systematic procedures to guarantee long-term financing. Nor could such mechanisms or procedures survive, unless based on widely shared assumptions about the objectives to be achieved or the rules to govern them. To the extent that an 'international society' can be identified, there may already exist a rudimentary point of departure for international discussion, but this has yet to be reflected in the creation of an accepted locus of decision-making for aggregated financial decisions.[5]

Gradually the international system has acquired in embryo elements of collective endeavour to mobilise and deploy limited but not insignificant financial resources. International organisations possess funds, though they consist primarily of administrative budgets and restricted expenditure programmes. Even if these are quite important within their own fields of operation, they do not constitute active or dynamic budgets capable of expressing political or economic priorities. Nor is this their object, since their operations are confined to the specific allocation of funds for limited purposes. There is, however, another feature of the international system that is beginning to provoke change, in that the heightened awareness within nation states of economic interdependence is stimulating debate about the responses appropriate to this predicament. This is reflected in attempts by groups of countries at least to co-ordinate elements of national economic policies, which have some impact on national budgets. Such has been the case, for example, among the members of the Organisation for Economic Co-operation and Development (OECD) or within regionally based international organisations such as Comecon or the Andean Pact. This trend stands witness to the impact of individual countries' budgets upon each other, as the scope for autonomous national policies diminishes.

Community Finances

Where then do the European Communities (EC) fit into this spectrum? Are their financial activities such as to warrant comparison with national budgets? Or do they simply offer a developed example of the concerted but limited endeavour that characterises international co-operation more generally? Moreover, from an analytical point of view it is of interest to determine whether the case of the EC offers any new insights into the theoretical debate that has been engaged among economists and political scientists about 'fiscal federalism' or the dispersion of financial powers among different levels of government.[6] A major factor in any discussion of the EC stems from the aspirations of some of their founders to establish a new and supranational government in Western Europe, rather than simply yet another international organisation. An essential activity of government is its budgetary process, and a vital attribute of a political system is its capacity to resolve controversy over revenue-raising and expenditure. It follows that a necessary condition for political and economic integration to occur is the acquisition of a fully fledged budgetary process, one in which significant financial resources are mobilised and which enables choices to be made about expenditure and the joint promotion of articulated policies.

On the face of it the evidence that emerges from the experience of the EC is ambiguous. Community financial activities are quite extensive; revenue is raised directly for the Community exchequer; and a variety of different expenditure programmes channel cash to the clients of EC policies. The sums of money are significant, if not enormous; the 1979 budget amounted to £8,750 million, equivalent to 0·8 per cent of Community GDP or 2·6 per cent of the sums spent in the national budgets of member states. The basic framework for this lies in the Treaties, but since then the scope and coverage of Community expenditure has steadily expanded. In parallel Community institutions have acquired additional authority and responsibility in terms of both initial decision-making and executive powers. The majority of these financial instruments are incorporated within a formal 'core' budget, with an annual cycle that resembles, at least in outline, the budgetary processes of individual countries. The other instruments are managed separately by other EC institutions and agencies. This pattern of financing enables the EC to intervene within the member states to finance major policies, notably the Common Agricultural Policy (CAP), and to pursue a variety of other sectoral objectives that complement and supplement national programmes. The other side of the coin, however, indicates that there are very considerable limits to the impact of the EC through their financial activities. The sums of money involved remain small as a proportion of Community GDP or by comparison with rates of public expenditure within the member states. There is a statutory ceiling on the raising of revenue, and the current resource base is very inelastic compared with national fiscal resources. Apart from agriculture, and possibly overseas development assistance, the financing of Community policies according to Community criteria is slender in scope and clearly subsidiary to the parallel programmes of national expenditure. So far there is little room for choices over priorities or for switching expenditure from one sector to another as needs change or policies shift. Nor is the budget flexible enough to replace existing financial instruments easily with new types of intervention. The framework of decision-making within some Community institutions, notably the Council of Ministers, continues to express the separate interests of individual member states rather than the aggregation of Community-wide interests. National administrations and agencies provide the channels through which expenditure is disbursed in such a way that Community criteria often fail to guide precisely the allocation of resources. The parameters that thus circumscribe EC financing make it seem insubstantial and ineffectual, by comparison with unitary nation states and even with federal states.

Clearly, therefore, the record of the EC needs very close and careful examination before a general assessment of its significance can be risked. Here the student of the EC runs into major difficulties because of the sparseness of research on Community finances. Political scientists have paid little attention to the budgetary process apart from their accounts of the newly found influence of the European Parliament over budgetary decisions.[7] Studies of particular policy sectors have not dealt in detail with their financial implications, except in broad terms.[8] Contributions from the practitioners largely concentrate on the specifics of processes and procedures without exploring the trends and their implications.[9] Only recently have economists begun to grapple with assessing the impact of EC finances, and the analyses so far published scratch only at the surface, often in a rather controversial manner.[10] As with much of the literature on the EC, there is a preoccupation with what might or ought to be the case, rather than with careful explanation either of what actually happens or of the underlying patterns of behaviour and their implications.

The object of this study is to clarify the discussion of budgetary politics within the EC and to relate them to broader analytical approaches. Inevitably this requires some exposition of the basic features of EC finances as well as of the various interpretations of the empirical evidence. Chapter 2, 'Community Revenue and Expenditure', identifies the main sources of income that comprise the 'own resources' of the EC and the mobilisation of other kinds of finance. It surveys the various expenditure programmes with their different instruments of direct policy support, grants and loans. Chapter 3, 'The Decision-Making Process', outlines the procedures, structures and bargaining through which Community and national institutions determine the objectives and content of EC finances. Chapter 4, 'Management and Control', documents the implementation of expenditure programmes at both Community and national levels, and examines the controls – political, administrative and financial – to which they are subjected. The final chapter explores the general implications and some interesting current developments, and then identifies areas for further research.

The Issues

In setting the context for this investigation the main issues that surround the subject need identification. First, what is the role of common financing within the EC? Secondly, what is the link between EC financing and the economics of redistribution or convergence? Thirdly, what are the criteria that should govern national contributions into and receipts from common funds? Fourthly, what does the particular discussion of EC financing reveal about institutional relationships and capabilities? Fifthly, what light does the evolution of EC finances shed on the argument about the development of the EC into a viable level of government in its own right?

THE ROLE OF COMMON FINANCING

The treaty framework deliberately endowed the EC with the right to raise and spend money on a far more extensive basis than in other organisations for co-operation among nation states. It explicitly envisaged the creation of financial instruments that would permit both positive intervention and selective grants and loans. However, the scale of operation was limited in four ways. First, there was no direct commitment to shift responsibility for public finance from the national to the Community level, except in specified fields of operation, such as agriculture. Such resources as the EC could muster thus tended to coexist with national financing and to impinge on only part of national budgetary responsibilities. Secondly, the implication was that the EC would raise only as much revenue as was needed for the support of Community policies, rather than spend as much as could be borne by the providers of revenue. Thus the effect was necessarily to circumscribe the economic scope of Community financing. Thirdly, there was no room for the EC to run a deficit or to acquire debts in the manner that has become common at the national level, apart from in the context of the purely banking operations of the European Investment Bank and the loans raised, for example, by the High Authority of the European Coal and Steel Community. Fourthly, the main spending programmes were set out according to distinct and separate rules as independent operations, rather than as a collective whole, as the outcome of an incremental process of policy co-operation.

The consequence was that the basic conception of Community finance was of the mobilising of limited resources confined to the support of particular policies, and not of a general financial responsibility to be exercised collectively. To express this in more economic terms the intention was simply to allocate specific resources for specific objectives. There is a parallel here with the historical development of national budgets, but these latter have increasingly been engaged in making an explicit redistribution of resources, and in adopting measures for the general stabilisation of their economies. Elements of incidental resource transfer inevitably have crept into the EC budget, but as an unintended or secondary effect. Policies receiving financial support from the EC were initially intended to be applied throughout the member states rather than selectively, though their incidence might not be even, and only some – the European Social Fund (ESF), the European Investment Bank (EIB) and the Regional Development Fund (RDF) – deliberately favoured the poorer regions or countries.

Over the years this simple model has grown more complex as new expenditure programmes have been added, as expectations within the member states have altered and as Community funding has begun in practice to have a differential impact on particular member states or sectors. However, there remains a difficulty in reconciling theory and practice. The orthodox theory of a limited EC budget to support specific policy objectives is still embraced in many quarters. While there is now some recognition that expenditure policies should tend to help the poorer areas, there is not yet acceptance that the revenue-raising system has a part to play in supporting this principle. A further complication stems from the underlying philosophy of the whole Community experiment, which was broadly to endorse the principle of a free and competitive market, with interventionist policies only for exceptional cases such as coal or agriculture. This philosophy also implies a limited budget, rather than a progressive increase in public finance. By contrast a more *dirigiste* approach to economic management across the board would require as a corollary an expanding budget from which to fund appropriate policies on a regional or sectoral basis. Clearly this would demand a qualitative change in the approach to EC finance. Whether that can or should occur is a question to which we shall return later.

REDISTRIBUTION AND CONVERGENCE

The narrow study of Community finances tells only part of the story of the effects of participation in the EC upon the member states' economies. The objective of the Treaties was to create a single market which might ultimately result in the emergence of a single integrated economy. The route through which this would be reached was by the dismantling of internal barriers and the stimulus that this would give to economic forces. These moves would constitute the 'dynamic' effects of Community policies in the form of a customs union, a commercial policy, an agricultural policy, common rules of competition and the removal of market distortions. A subsidiary arm would be the direct deployment of EC finances to back up some of these policies, but their effects would be rather more 'static' and rather less important. The conviction that the dynamic effects were the prime objective of economic co-operation in the EC meant that the budget was not initially viewed as directly providing instruments for economic management.[11] The financial activities directly engaged in by the EC might none the less have some spillover effect of a more dynamic kind on economic performance in member states and their regions, but only of a secondary nature.

As the EC have evolved, the sharpness of the distinction between dynamic and static economic effects has been eroded, partly because of the political interpretations within some member states of the impact of Community policies. Notably since the major recession of the mid-1970s commentators have increasingly drawn attention to the divergence between the economically stronger and the weaker member states in terms of various economic indicators such as relative national wealth or price and wage levels. This is reflected in the explicit differentiation between the 'prosperous' and the 'less prosperous' countries that has increasingly since 1978 been drawn into the debate about new policy initiatives within the EC and especially over the new European Monetary System (EMS). In this context it is important to recall that one of the purposes written directly into the Treaty of Rome was:

> The Community shall have as its task, by establishing a common market and progressively approximating the economic policies of Member States, to promote throughout the Community a harmonious development of economic activities, a continuous and balanced expansion, an increase in stability, an accelerated raising of the standard of living and closer relations between the states belonging to it.[12]

The two assumptions that underpinned this were, first, that economic growth would continue at a substantial rate, and, secondly, that the economies of the member states would follow compatible trends fostered by the establishment primarily of the 'four freedoms' of movement for goods, services, capital and people. Only the economy of Italy at the time manifestly lagged behind those of the other member states, and recognition of this was reflected in the design of the ESF and the EIB.[13] These apart, no explicitly redistributive features for EC financing seemed necessary. On the contrary the basic philosophy applied to most policy sectors rested on the adoption of common criteria to guide EC financial intervention throughout the member states.

Over the years this philosophy has become more and more vulnerable to attack. As national policies have altered and as the international environment has become more turbulent, it has ceased to be possible to rely on sustained growth and stable economic performance. The clear differences among the member states' economies have looked more important than the common trends, even to the extent where some commentators now argue that there is definite evidence of increasing divergence. Most would now also accept that there is a striking incompatibility of economic interests and priorities among member states. There is fierce debate among economists over both the nature of this phenomenon and its relevance to the EC.[14] Three other factors have combined to accentuate its political repercussions. First, the 1973 enlargement brought into the EC two new members – Ireland and the UK – with levels of economic performance substantially below the Community average and akin only to that of Italy among the founder members.[15] Over six years later the record suggests that EC policies have positively helped the Irish economy, while they have not necessarily produced more benefits for the UK. Secondly, renewed interest in monetary co-operation and the establishment of the European Monetary System (EMS) have made more explicit the distinction between the 'prosperous' and the 'less prosperous' in so far as this is reflected in currency values.[16] This led the governments of the three 'less prosperous' countries to argue that their participation in the EMS would require specific EC measures to help them to catch up, though in the event the UK did not join the scheme at its inception. Thirdly, the Mediterranean enlargement of the EC, to bring in Greece in 1981 and Portugal and Spain at a later date, will swell the ranks of the 'less prosperous' and thus increase the extent and scale of divergence. One possible outcome envisaged in response is the emergence of a 'two-tier' or 'multi-velocity' Community in which some member states would more rapidly embark on further common economic policies than the others.[17]

What then is the relevance of this debate to the EC budget? A major feature of national budgets is their acceptance of responsibility for selective revenue-raising and financing, either to equalise income in the different parts of the country or to redistribute wealth. This is true of both unitary and federal systems, though with the important distinction that in unitary systems the transfers of resources are often hidden while in federal states they generally depend on difficult and explicit political negotiation.[18] The scope and nature of the budgetary mechanisms vary enormously from the highly sophisticated to the rudimentary, but the political case is accepted for using public finance to offset some of the effects of differential economic performances. This is not to say that national budgets can or necessarily should seek to eliminate disparities, but they cannot escape involvement in mitigating the consequences of economic disparities.

If the EC are in the business of promoting harmonious development and balanced expansion of their economies and also engaged in increasing their political cohesion, then the issue of redistribution and resource transfer cannot be avoided. The words 'harmonious' and 'balanced' are ambiguous, since they leave open the questions of whether harmonious implies equal or complementary, and of whether balanced applies to comparisons among countries or among regions. But the appearance of these words in a constitutional document in itself generates expectations of a Community commitment, unless the phrases are to be dismissed as purely symbolic. Moreover the issue becomes more acute, if, as some would argue, the actual consequence of some EC policies is to reinforce the secular trends of divergence. If the impact of EC policies is not neutral, then the Community process is already engaged in the redistribution of resources, and the question which follows is how far deliberate measures can or should be undertaken at a Community level to benefit selectively the 'less prosperous' countries and regions. Here there is a major difference of opinion between those who see this as clearly a Community responsibility and those who argue that, whatever may be the role of the EC, it is and must remain relatively insignificant beside the responsibility of national governments, unless the EC were to change fundamentally and become a developed federation. If Community responsibility were generally accepted, then the EC budget and other Community instruments presumably would constitute an important vehicle through which to transfer resources.

A further complication arises from the way in which the EC have acquired new members. The original blueprint was based on a compact among only six countries in the 1950s, a compact which identified a hierarchy of political objectives to meet their priorities as then perceived. The compact was entrenched in the Treaties and reinforced by successive negotiations during the 1960s. But new members bring to the EC not just endorsement of the *acquis communautaire* and support for more of the same, but different pressures and needs that require different policy responses. This was true of the first enlargement, notably in the case of the UK, and will equally be a consequence of Mediterranean enlargement to include Greece, Portugal and Spain.

The implications of this are twofold. In the first place new members expect to be able to call on Community resources to meet their domestic needs, expectations inevitably expressed in specific claims on EC finances. The 1973 enlargement produced the impetus for the creation of the Regional Development Fund (RDF), which had a built-in element of redistribution in the quotas apportioned to individual member states. The prospect of Mediterranean enlargement is already stimulating demands for other kinds of selective policy innovation.[19] Secondly, the policy areas currently central in EC debate are no longer confined to the relatively distinct and in some ways rather special sectors such as agriculture. They include industrial, regional and energy policies, for example, which are different in kind in that, while all member states have an active interest in these sectors, common policies could not be applied uniformly except at very great financial cost, or with extensive EC intervention into national economic policies. In any case some member states at least would prefer a much more selective approach to Community intervention. Lastly on this point it should be noted that the concern about economic divergence reflects in part a dissatisfaction on the part of some member states with the current budgetary priorities of the EC as much as an expectation that large-scale resource transfers are feasible through EC mechanisms, or that EC policies themselves affect divergent economic performance.

CRITERIA FOR NATIONAL CONTRIBUTIONS AND RECEIPTS

The Community experiment was predicated on the assumption that all member states would derive sufficient benefit from participation to outweigh the costs. In EC negotiations, as elsewhere, success depends on a blending of the gains and penalties incurred sufficiently attractive to all the parties to permit a consensus. While this may be a desirable approach to bargaining its application to the budgetary process is fraught with problems. All parties can gain significantly only if the overall resource base increases substantially and permits some to win a higher proportion of the resources than others. Even to preserve a consensus this far requires delicate and sensitive judgements about financial allocations, in order to ensure that those who in practice contribute heavily perceive this as a worthwhile investment in order to reap other non-financial returns. If, however, resources are diminishing or growing only marginally there is automatically much sharper conflict about the balance-sheet of credit and debit of each participant. If in addition there is dissonance between the net financial position of particular countries *vis-à-vis* the EC budget and their shares of Community economic wealth, serious controversy is unavoidable.

In the early years of the EC this controversy was not directly exposed, since the overall trade-offs endorsed by the participating states covered a broad spectrum of economic and political interests. Amongst these, financial instruments had a relatively low priority and were on the whole subsidiary to other objectives. Two elements of the compact did, however, have an explicitly financial character. Some instruments – the EIB and the ESF – were specifically included to help to serve the interests of Italy, clearly the poorest member state. Secondly, the agreement to include a common agricultural policy and a programme of overseas aid carried a commitment to substantial Community financing, in response primarily to French insistence on their inclusion. Member states viewed these, as it were, as a necessary subscription to establish the club, but without considering fully their long-term implications.

Confidence in the continued rapid growth of their economies combined with the expectation that agricultural producers would make diminishing claims on the EC resources to obscure the ramifications of these initial commitments. Notwithstanding this, from the outset an important strand in French policy was that the financial gains of EC membership should at least balance the financial contributions – the so-called principle of *juste retour*.[20] Arguments during the 1960s threatened from time to time to disturb this consensus, but it was not until the end of the decade that it came into the open. As the costs of the CAP rose, concern was voiced in West Germany at the high burden imposed by a net German contribution to the EC budget. From then on pressures to keep down the size of the budget became a major constraint on the development of new expenditure programmes, reinforced by a desire to avoid the open-ended commitment to EC finance for new areas that might repeat the experience of the CAP. These pressures were particularly evident in the long controversy of 1973–5 over the establishment of the RDF.[21] They reflected the extent to which perceptions of unfairness and a sense of grievance had come to impinge on the debate about both new EC policies and their financial implications.

Until the 1973 enlargement two member states apparently suffered from the structure of the EC budget: West Germany because of its high contributions, and Italy because of its limited receipts. The continued expansion of the German economy made the German case for change look weak to partner governments, while other incidental factors seemed to explain the low Italian take-up of EC funds, ranging from administrative problems to the difficulties of enacting appropriate implementing legislation. In 1973, however, the context of the debate changed completely, though this became evident only gradually. The accession of the UK brought into the EC a member state in which specific concern about the impact of the EC budget had already been a major factor in domestic political debate. In 1970–2 estimates of an anticipated net contribution ranged between the adverse and the very adverse.[22] Given the broader political controversy in the UK over the principle of EC membership the forecasts could easily be dismissed elsewhere in the EC as prognostications of doom by anti-marketeers. The Treaty of Accession contained in Articles 129–31 provisions that staggered the impact of British contributions, so that the full picture was not immediately visible. During the 'renegotiations' conducted by the Labour government in 1974–5 a further limit to the UK contributions was added in the form of the Financial Mechanism, although it was not in practice to be activated until 1979–80.[23]

These provisions led to the widespread assumption in the EC that the problem had been largely overcome, while in the UK hopes of increased budgetary receipts to compensate softened the impact of the contributions. But during 1978 it became crystal clear that the problem had far deeper roots. Figures began to circulate indicating the substantial rate of increase in British contributions, and by spring 1979 it was evident that the UK was overtaking West Germany as the single largest net contributor to the EC budget. This situation arose in spite of the UK's low position in seventh place in the league table of EC economic performance, with Italy the victim of a similar fate. Conversely some member states with rather more favourable economic circumstances, notably Denmark, were deriving significant cash gains through the EC budget.[24] Tables 1.1(a) and 1.1(b) illustrate this.

Inevitably hard political argument has accompanied the exposure of this problem. Underlying this is a dispute about the application to EC mechanisms of the principle of equity. While no one pretends that the EC budget represents a complete picture of

Table 1.1   *Net Transfers Through the Community Budget*
(£m.)

(a)   *Derived from EC Commission figures*

	1976	*1977*	*1978*	*1978 (after full Article 131 refunds)*
Belgium-Luxembourg	+ 222·3	+ 247·4	+ 261·2	+ 252·6
Denmark	+ 235·5	+ 339·7	+ 411·9	+ 411·9
Germany	− 630·8	− 844·4	− 230·3	− 281·4
France	+ 63·5	− 30·4	− 22·1	− 55·0
Ireland	+ 120·8	+ 267·0	+ 352·2	+ 356·0
Italy	+ 130·4	− 43·4	− 480·3	− ·5
Netherlands	+ 183·4	+ 187·0	+ 157·2	+ 146·4
United Kingdom	− 148·0	− 408·0	− 744·6	− 625·8
TOTAL	+ 177·1	− 285·1	− 294·8	− 294·8

Source: Written answer to a parliamentary question in the British House of Commons, HC *Hansard*, 12 June 1979, col. 141, based on a press release published by the Commission.

Notes:
1   These are net recorded budgetary transfers, and are not adjusted to express MCAs as benefits to countries other than those in which the actual cash payments are made.
2   The sign + = net beneficiary; − = net contributor.
3   The final column shows the eventual position for 1978 when, under a decision of December 1977, the provisions of Article 131 of the Treaty of Accession were met for the new members by repayments in arrears of excess contributions.

(b)   *An Alternative View for 1978*

	Net budget receipt	Net trade receipt	Total net cash receipt
United Kingdom	− 806	− 317	− 1123
Germany	− 570	− 101	− 671
Italy	− 114	− 532	− 646
Belgium–Luxembourg	+ 312	− 156	+ 156
Ireland	+ 254	+ 221	+ 475
Netherlands	+ 190	+ 441	+ 631
Denmark	+ 329	+ 289	+ 618
France	+ 114	+ 620	+ 734

Source: Cambridge Economic Policy Review, April 1979.

Notes:
1   These figures are based on earlier Commission estimates of actual transfers, hence the discrepancies with Table 1.1(a).
2   The 'net trade receipt' column is based on an estimate of the difference between the food costs that result from EC policies and the hypothetical cost of food at 'world prices'. The effect, according to the Cambridge Economic Policy Group, is to increase the actual cash benefit of EC membership to the main food-producing countries and to increase the cost to food-importing countries.

the economic effects of membership on the participating states, contributions to and receipts from it are visible and identifiable. The budgetary flows to some member states have been shown to be perverse in terms of the disproportion of individual member states' net positions. The asymmetrical relationship between cash flows and comparative economic performance compounds the unfairness and sense of grievance. Other dimensions of the continuing debate over EC membership within the UK mean that the budgetary issue has high political salience, both symbolically and because of its substantive repercussions. Consequently it impinges on debate within the EC over both existing policies and new proposals. Even in Italy, where the pro-European consensus remains strong, similar themes are now emerging in domestic debate.

How then should the principle of equity be defined in order to provide criteria against which to determine contributions to and receipts from the EC budget? Here both economists and political scientists have contributions to make. The principle of equity is deeply embedded in the traditional approaches of economists to public finance.[25] There are, however, two different strands to the interpretation of this doctrine. The first, horizontal equity, is that all individuals or groups that fall into a particular category should be taxed similarly. Thus for example, in the UK, local rates are levied according to the dwellings or premises occupied, and particular kinds of economic transactions are taxed at uniform levels. Logically, too, expenditure grants are disbursed evenly to all groups or individuals who meet certain defined conditions. The assumption is that the criteria for both revenue and expenditure are determined objectively, and that individuals or groups have the choice as to whether they behave in such a way as to meet the criteria. The second strand, vertical equity is that where there are discrepancies in the positions of individuals or groups depending, for instance, on their location, level of income or their costs, distinctions should be made in their tax burden and their qualifications for budgetary receipts. Consequently the deployment of public finance must be rather more selective and discriminating. From this flow the practices of incorporating criteria of progressivity in revenue-raising, namely, that tax burdens should be related to the ability to pay, and of using equalising measures, either to ensure minimum provision of services or to compensate for disadvantage or, more radically, to redistribute wealth. National budgets in general apply both strands of the doctrine to their different financial instruments. Notionally at least this is not too difficult in a centralised financial system, a unitary state, where the instruments are managed by a single set of institutions. Obviously it is far more complex in a decentralised system, such as a federation where both authority and instruments are dispersed. There is now a considerable literature on the implications of this for 'fiscal federalism'.[26]

If we look at the record of the EC it becomes rapidly evident that so far only the first strand of the economic doctrine is explicit in Community practice. The revenue base of 'own resources' concentrates on sources of income that emphasise the uniform application of specified criteria. Similarly the major share of expenditure for Community programmes is apportioned to all those who meet particular conditions. Only a limited share of Community finance is disbursed selectively according to complex criteria of eligibility or priority, and only recently has the concept of assessing revenue according to ability to pay begun to creep into the discussion.

The approach of political science to the definition of equity is less easily summarised since it stems from an admixture of very different elements. There is a body of literature on political philosophy that has sought to clarify fairness and equity in a polity, though more often as applied to rules of justice and individual rights than to social and economic activities. Various frameworks for the analysis of the political systems have stressed the importance of establishing processes through which conflict over the distribution of public goods can be resolved in such a way that individual and group interests are protected by collective action according to criteria of fairness. Others have looked at the linkages between economics and politics and endeavoured to define the application of political criteria to economic activities and the constraints imposed on politics by economic developments. Lastly, there is an extensive literature about the particular political difficulties that arise in the determination of policy choices and resource allocation in multi-level systems of government, often in the specific context of federations.[27]

Without going into the details of all these areas of discussion among political scientists, three major points emerge. First, in practice whether a political system is stable depends in large part on whether its operations are perceived as fair by at least a substantial majority of its members. If even a small minority perceives its position and treatment as unfair this may create serious conflict for the political system as a whole. Subjective interpretations of fairness and unfairness carry great influence irrespective of attempts to establish universal and objective rules. Secondly, politics is a continuous process of conflict resolution based on permanent 'renegotiation' of outcomes according to authoritative procedures. As the environment of the system changes, and as members respond to changing needs and circumstances, new demands are continually fed into the authorities which often require either new policies or the redefinition of old policies. Consensus and acceptance of the outcomes depend on the capacity of the authorities to respond to these changing pressures. Changing claims on public financial resources illustrate this very clearly. Thirdly, in a system which possesses multiple levels of authority the process through which political bargaining takes place is highly complex and has built-in tensions that render the achievement of consensus and the preservation of fairness extremely difficult.

The relevance of this to the EC is considerable. First, it

emphasises the significance that attaches to perceptions of unfairness. To the German it has seemed unfair that his contribution to the Community exchequer is high, while to others it may seem only proper that a wealthy German should contribute a lot more than a poor Italian and quite a bit more than a modestly affluent Frenchman. To a British consumer it appears indefensible to pay heavily to the Community budget in order that a German or Danish farmer should have a rather high income, while in retort it can be argued that it is quite fair for this to happen if the explanation lies in the continued preference of the British for importing foods from outside the EC. Secondly, it highlights the importance of establishing a framework for resolving conflict over finance that is flexible enough to adapt to a changing environment. If policies reflect what might once have been equitable criteria that have since become disputed as a consequence of changing pressures, then those who suffer from the application of the criteria will inevitably regard the situation as unjust. This in turn diminishes the stability of the system and reduces respect for the procedures through which decisions are taken. Thirdly, it is quite clear that the EC offer an example of multi-level decision-making which itself aggravates the problem and makes the resolution of conflict according to principles of fairness particularly hard. Lastly we should note that the debate about equity is important irrespective of whether the EC develops into a tightly knit and cohesive political system. The EC could not become an accepted locus of decision clearly superior to member states unless these latter were individually confident that the rules of the game would operate fairly. However, even to preserve the rather looser association that currently constitutes Community co-operation requires the conviction in all participating states that fair criteria govern decisions about financial contributions and receipts. If the criteria are perceived as inequitable, then this in turn is likely to reduce support for Community authority, and weaken the prospects for its further extension.

INSTITUTIONAL RELATIONSHIPS AND CAPACITIES

Any empirical example of Community activity helps to clarify the institutional processes of the EC and to explain the political interactions that they embody. Much of the recent literature on the EC has consisted of separate studies of individual issue areas and policies, often leading to the conclusion that different processes can be identified depending on different contexts such that grand generalisation about the overall pattern of decision-making is inapt. The application of this approach to EC finances would seem to offer particularly fertile ground to the researcher who seeks to add to an understanding of the Community phenomenon. The budgetary process has the advantage of following a fairly systematic annual cycle that involves all the major EC institutions (except so far the Court of Justice), the national governments of member states, and to some extent regional and local authorities. This affords a breadth of institutional coverage and the opportunity to compare trends and patterns from year to year. The process has an identifiable output that can be examined in detail, unlike some aspects of EC activity where the outcomes are less tangible. Though the budgetary process does not represent the sum total of EC activity, it does bite into many of the major policy areas and thus sheds light on both individual policies and the relationships among them. Because the EC have had financial instruments and a budget from their initial establishment it is possible to discern from them features of the evolution of the EC throughout their history, with one extra bonus to the researcher that major innovations in both substance and procedure offer scope for the analysis of explicit change as well as implicit shifts in processes.

More specifically the example of EC finance provides an opportunity to look in detail at both the internal functioning of the individual institutions involved at Community and national levels and their capacity to handle their component of the overall process. Equally it illustrates important features of the relationships among these various components. Three different dimensions have to be incorporated into the analysis; the political, the administrative and the more narrowly financial. Lurking behind these are the broader issues of accountability and control, efficiency of management and the effectiveness of resource deployment. Here the link must be made between assessments of the performance so far of the Community process in this important area of activity and views on the acceptability of extending co-operation either into new fields or to increase the general authority of EC institutions. If the Community process can be shown to have performed well on these different dimensions then this might increase support for strengthened collaboration. But should the record look more doubtful this might either induce a reluctance to proceed further or perhaps increase pressures for institutional changes to meet criticism. It is also important to bear in mind the parallels to be drawn with the performance of individual governments in dealing with their own budgets, since those from member governments who participate in Community negotiations or who administer EC finances at the national level inevitably assess the record of the EC against a backcloth of their own national traditions and practices. Since these vary enormously from one member state to another, this constant implicit comparison helps to explain the very different perceptions and expectations of the role of EC financial activities. These questions will be pursued further in Chapters 3 and 4.

POLITICAL COHESION AND FRAGMENTATION

In discussions of the politics of the EC the budgetary issue has figured particularly in assessments of the future role of the European Parliament. The opportunity that the Parliament now enjoys to exercise a significant degree of influence over appropriations offers the scope for an increase in democracy within the EC process, which otherwise has rested indirectly on the answerability of individual member governments to their own electorates. Already, some have argued, there has been a shift of emphasis, if not of power, to the European Parliament, as it has increasingly interfered in the bargaining among governments and with the Commission, by questioning expenditure and even altering the figures attached to particular programmes. The question which then follows is how far the budgetary arena will enable the Parliament to increase pressures for constitutional change in the EC.

However, underlying the analysis of Community finances is the broadest question of all, namely, its relevance to any assessment of the degree of political cohesion in the EC. On the one hand, the politics of the budget reveal both the pressures for and constraints on cohesion among the member states. This is especially illuminating in that successful negotiations about finance require at least some of the participants to put, as it were, their money where their mouths are, rather than to limit their commitment to EC co-operation to the declaratory. Also it can be argued that enthusiasm on the part of governments for Community co-operation is not fully tested if their benefits are significant in cash terms. Equally interesting is the question of how far the EC are moving towards more tightly knit collaboration with aggregated decision-making power. The main hypothesis to be tested here is whether a larger and more sophisticated budget can emerge only if a greater degree of political cohesion develops, sufficient to allow a consensus on both the objectives and the instruments of common policies. Among the indicators of such a shift would be: first, the extent to which active choices could be made through the budget; secondly, the number of economic levers that would need to be manipulated through a budget at the EC level; and thirdly, the degree to which member states would accept not just money from EC coffers but also guidance about how it should be spent.

For a budget to do more than maintain current programmes demands consensus on new objectives and on priorities. If the EC budget is to do more than allocate finance to a restricted number of specific policies and to engage, for instance, in counter-cyclical measures, governments would have to accept that it would take over some of the functions that currently are located at the national level. The prospects for a significantly larger total budget seem closely linked to the issue of what strings would be attached by Community institutions to the deployment of EC resources in member states, especially if this were to rest on explicitly redistributive instruments.

Notes: Chapter 1

1 See, for example, W. J. H. Mackenzie, *Politics and Social Science* Harmondsworth, Penguin, 1967, pp. 137–52; or Robert Dahl and Charles E. Lindblom, *Politics, Economics and Welfare*, New York, Harper, 1953.

2 See P. Samuelson, *Economics*, New York, McGraw-Hill, 5th edn, 1961, ch. 19; C. T. Sandford, *The Economics of Public Finance*, Oxford, Pergamon, 1977; and Andrew Shonfield, *Modern Capitalism*, London, Oxford University Press, 1965.

3 For a general comment see P. Self, *Administrative Theories and Policies*, London, Allen & Unwin, 1972, pp. 261–77.

4 See, in particular, James E. Danziger, *Making Budgets*, London, Sage Library of Social Research, 1978; and Aaron Wildavsky, *Budgeting: A Comparative Theory of Budgetary Processes*, Boston, Mass. Little, Brown, 1975.

5 See Michael Donelan (ed.), *The Reason of States: A Study of International Political Theory*, London, Allen & Unwin, 1978, especially Christopher Brewin's essay 'Justice in international relations', pp. 142–52.

6 For a general discussion see Wallace A. Oates, *The Political Economy of Fiscal Federalism*, Lexington, Mass., Lexington Books, 1977; and for a series of

comparative studies see *The Role of Public Finance in the European Communities*, 2 vols, Brussels, Commission of the European Communities, April 1977 (hereafter the MacDougall Report).

7   For a general discussion see David Coombes (ed.), *The Power of the Purse*, London, Allen & Unwin, 1975; and for a more specific commentary on the EC, David Coombes and Ilka Wiebecke, *The Power of the Purse in the European Communities*, London, RIIA/PEP, 1972.

8   For example, Doreen Collins, *The European Communities: The Social Policy of the First Phase*, 2 vols, London, Martin Robertson, 1975; Carol Cosgrove Twitchett, *Europe and Africa: From Association to Partnership*, Farnborough, Saxon House, 1978, esp. pp. 155–63; and Ross Talbot, 'The European Community's Regional Development Fund', *Progress in Planning*, vol. 8, no. 3, 1977.

9   The most comprehensive account is provided in Daniel Strasser, *The Finances of Europe*, New York, Praeger, 1977, with further points covered in the same author's periodic articles in *Revue du Marché Commun*, Paris.

10   See *The MacDougall Report; The Role of Inter-regional Flows of Public Finance in the European Community*, Brussels, Groupes d'Etudes Politiques Européennes, Report No. 3, 1979; Geoffrey Denton, 'Reflections on fiscal federalism', *Journal of Common Market Studies*, vol. XVI, no. 4, June 1978, pp. 283–301; and Wynne Godley, 'Policies of the EEC', *Cambridge Economic Review*, April 1979, pp. 23–30.

11   For a review of the literature see Jacques L. Pelkmans, 'Economic theory and the impact of the Customs Union on the Common Market', unpublished paper, University of Tilburg, June 1979. See also Bela Belassa (ed.), *European Economic Integration*, Amsterdam, North Holland, 1975; and Sir Alec Cairncross *et al.*, *Economic Policy for the European Community: The Way Forward*, London, Macmillan, 1974.

12   Treaty establishing the European Economic Community, Article 2. See also the reference in the preamble (para. 5) to the need to ensure 'harmonious development' and to reduce differences among regions.

13   *Report from the Delegation Heads to the Ministers of Foreign Affairs* (the Spaak Committee report), Brussels, 1956.

14   See Michael Hodges (ed.), *Economic Divergence in the European Community* (provisional title), London, Allen & Unwin, forthcoming.

15   See Table 2.7, p. 63.

16   These terms were coined in autumn 1978 during the 'concurrent studies' exercise, in which member governments discussed, through the Economic Policy Committee of the EC, the case for economic measures in parallel to EMS. The 'less prosperous' category was defined initially to include Ireland, Italy and the UK which were to be eligible for special help in the form of interest rebates on EC loans. In the summer of 1979 there was continuing disagreement as to whether the state of the British economy justified its continuing categorisation as 'less prosperous'.

17   See the special issue of *Europa Archiv*, August 1978, on enlargement; and Michael Hodges, 'The legacy of the Treaty of Rome: a Community of equals?', *World Today*, June 1979.

18   See Oates, op. cit.; the MacDougall Report, Vol. II; and Wildavsky, op. cit., esp. chs. 2 and 3.

19   On the RDF see H. Wallace, W. Wallace and C. Webb (eds), *Policy-Making in the European Communities*, Chichester, Wiley, 1977, ch. 6; and on the Mediterranean package see *Thirty-Third Report of the Select Committee on the European Communities of the House of Lords*, Session 1977/8.

20   The term *juste retour* means in theory a fair return. For the arguments put by the

French government in 1965–6 see Stephen Holt, *The Common Market: The Conflict of Theory and Practice*, London, Hamish Hamilton, 1967; John Newhouse, *Collision in Brussels*, London, Faber, 1968; and Joël Rideau, *La France et les Communautés Européennes*, Paris, LGDJ, 1975, pp. 365–78.

21   See Wallace, Wallace and Webb (eds), op. cit., ch. 6; and Ross Talbot, op. cit.

22   For the official British government view see *Britain and the European Communities; An Economic Assessment*, Cmnd 4289, London, HMSO, February 1970, and *The United Kingdom and the European Communities*, Cmnd 4715, London, HMSO, July 1971; and for an account of the broader debate see Simon Z. Young, *The Terms of Entry*, London, Heinemann, 1973, ch. 2.

23   See below, p. 59.

24   See Table 2.3, p. 51, and Table 2.7, p. 63.

25   See, for example, Oates, op. cit., ch. 1, pp. 12–14; and C. T. Sandford, op. cit., pp. 82–3.

26   See especially Oates, op. cit.; and the MacDougall Report.

27   See Danziger, op. cit.; Kenneth Hanf and Fritz Scharpf, *Interorganisational Policy-Making*, London, Sage, 1978, esp. ch. 13; and Wildavsky, op. cit.

2 *Community Revenue and Expenditure*

From an economic perspective three major issues are raised by the development of direct EC financing. First, what impact does it have on economic behaviour and performance in the member states and their regions? Secondly, what scale of resources would need to be drawn in if the EC were to establish collective instruments for macroeconomic management? Thirdly, which financial instruments should most appropriately be developed at the EC rather than the national level? All these issues have preoccupied economists in recent years. Some studies have attempted to measure and analyse the economic impact of EC finances, though they do not yet provide a clear or objective account.[1] Other studies have been more concerned with examining examples of multi-level financing in order to identify criteria that might be applied to the future evolution of EC finances. It is to this latter field that the Commission has paid particular attention, notably through the establishment of the committee of experts, chaired by Sir Donald MacDougall, on the role of public finance in European integration, to which we shall return later.[2] There is obviously an important distinction to be drawn between these two sorts of inquiry. The first examines the consequences of current activities, while the second rests on the assumption that a more active EC budget is in principle desirable. The link between the two is that the viability of an explicitly dynamic EC budget depends in part on whether economic analysts and politicians judge that the initial attempts of the EC to mobilise financial resources directly have been sufficiently effective to merit their extension.

It must, however, be recognised that an understanding of these economic issues cannot be derived solely from the narrow study of the direct financial activities of the EC. Other policy activities also impinge fairly directly, in that national budgets are affected and the general availability of finance is altered. The EC, for example, have attempted since the early 1960s to co-ordinate the economic activities of the member governments, notably for medium-term planning, short-term policies and national budgetary policies. Most commentators argue that the record of achievement is limited, in that the efforts of co-ordination have rested on intensive consultations rather than produced identifiable agreements of a formal nature.[3] However, the habit of consultation combined with the recognition that individual national budgets affect each other has led to a certain degree of synchronised action by national authorities, particularly as regards short-term counter-cyclical and stabilisation measures. This therefore constitutes a degree of Community intrusion into the budgetary processes of the member states in areas that in the views of at least some participants might appropriately be actually developed as a prime field of Community involvement. In addition it is important to bear in mind that there is a linkage between EC activities and the operation of the financial markets in Western Europe. Community institutions raise capital partly through the markets and to some extent recycle money back through the markets.[4] Also, EC legislation has begun to cover the operations of financial institutions in order to remove distortions to the market.[5] While these questions fall outside the boundaries of this study, their relevance should be recorded, although their influence as yet is fairly marginal.

In examining the direct financial activities of the EC, we should bear three general points in mind. First, there is a distinction between the formal 'core' budget of the EC and the other financial instruments that complement and supplement it. The budget itself includes most of the direct expenditure of EC institutions, apart from that of the ECSC with its special rules, and the EDF, though proposals have been made to incorporate this latter too within the common budgetary framework.[6] Most of the commentary throughout this study bears on the budget itself rather than on the other financial instruments. Secondly, an initial account must be given of the current scale of the EC budget, so that at least a preliminary impression can be formed of its significance. Table 2.1 shows the increase in the size of the budget, together with its share of Community GDP and its proportion of total public expenditure in the member states. The

Table 2.1 Expansion of the General Budget and Comparisons with GDP and National Budgets (EUAm.)

Year	Total	EAGGF guarantee as %	Budgets of the central governments of the member states	GDP of EC	General Budget as % of national budgets	General Budget as % of GDP
1973	4,641	77.4	227,800	870,200	2.0	0.53
1974	5,037	67.3	276,500	987,900	1.8	0.51
1975	6,214	69.6	351,900	1,132,600	1.8	0.55
1976	7,993	71.8	398,300	1,315,100	2.0	0.61
1977	8,483	76.8	448,300	1,483,800	1.9	0.57
1978[1]	12,363	70.2	471,300	1,546,800[4]	2.6	0.80
1979	13,716	69.9[3]	525,800[4]	1,716,900	2.6	0.80
1980[2]	14,997	60.8[3]	n.a.	1,895,500[4]	n.a.	0.79

Source: Adapted from the Preliminary Draft of the General Budget … 1980, Brussels, Commission, June 1979, Vol. 7/A, p. 109.

Notes:
1 Figures up to 1977 were originally in UA and have been adapted for EUA. From 1978 they represent EUA at average rates per year.
2 Figures for EC budget represent the Commission's proposals for 1980.
3 Includes refunds for food aid and certain sugar exports, which for 1980 the Commission proposed to categorise separately.
4 Estimate.

volume has steadily increased and at a faster rate than was anticipated either by the Commission or by member governments. Currently the sums of money involved still look small by comparison with the levels of public expenditure disbursed nationally or even by reference to the interventions of central governments within federal systems. But this relative modesty does not deprive the EC budget of financial and economic significance. Payments between member states and the EC cross the exchanges and thus affect their balance of payments and alter the availability of financial resources to national governments, quite apart from any broader economic effects. Thirdly, the EC budget has become technically very complex, partly because of the intricate rules that govern both revenue and expenditure, and partly because of the difficulties of establishing appropriate values for finances that are deployed in countries that preserve independent currencies, the mutual relationships of which are not static.[7]

Community Expenditure

Broadly speaking EC expenditure divides into three categories.[8] First, there is direct financing of distinctively Community policies. The best example of this is the guarantee section of the EAGGF, which finances in its entirety the common price policy of the CAP. Others include the European Development Fund, the arrangements for some direct Community aid either to non-associated developing countries or to countries which have special agreements with the EC, or the funding of specific EC projects, such as the Joint European Torus (JET) in nuclear energy research. Secondly, EC resources provide once and for all grants of various kinds through which Community criteria are superimposed on, or coexist with, national programmes. These include, for example, the operations of the ESF, the RDF and the guidance section of the EAGGF, part of the interventions of the ECSC, and a number of smaller programmes. Thirdly, the various loan facilities allow Community aims to be pursued through the promotion of particular kinds of investment, again either within member states or in other countries that are linked to the EC in some way.

Each programme of expenditure is based on Treaty provisions and subsequent legislation, and each depends on defined rules about its allocation and the criteria of eligibility. These rules differ significantly from each other, reflecting different negotiations at different times. Moreover, the formal structure of financing, especially within the General Budget, rigorously separates each type of expenditure into an individual compartment. Consequently the pattern of expenditure is rather fragmented, such that it is difficult to switch money from one heading to another as needs and priorities alter. Mechanisms do exist for transferring resources within a given financial year from one programme to another, but these are difficult to use. More important, it is difficult to alter the balance of expenditure among programmes from one year to another. This reflects the fact that total EC expenditure is the sum of individual component parts, each with a life of its own.

There is a parallel to be drawn here with national budgets, which similarly represent a mixture of different commitments embraced at different points in time, some of which are in practice difficult to alter. However, there are two crucial differences. First, to change or to reduce an expenditure programme in the EC is as difficult as to negotiate its creation. Secondly, EC finances lack the instruments of fine tuning that have come to comprise a significant element in sophisticated national budgets. Thus the operation of EC spending tends to be rather heavy-handed and slow to evolve. Programmes contain a composite package of elements designed to meet the claims and priorities of individual member states, which often obscures the general policy objectives that lie behind their introduction. Equally it is difficult and indeed misleading to try to identify from EC expenditure programmes and their relative size a broader philosophy or statement of policy priorities. The breakdown of the General Budget among sectors combined with the dates at which individual chapters were included illuminates the history and evolution of Community bargaining, as much as it indicates an economic rationale. Figure 2.1 summarises the main components in the current General Budget. This makes evident the differential 'success' of the EC in winning the consent of governments to Community intervention. The early consensus on agriculture is matched by the substantial share of resources devoted to the CAP, while the continuing disagreements on, for example, industrial policy are mirrored in the low volume of EC funding. Also it is difficult to envisage 'automatic' funding for industry on the model of the CAP. Each attempt to incorporate a new financial instrument requires prior agreement on the case for

Fig 2.1 Distribution of expenditure from the General Budget by policies.

Total appropriations for commitments in 1979

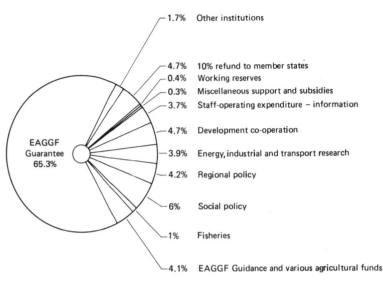

1.7%	Other institutions
4.7%	10% refund to member states
0.4%	Working reserves
0.3%	Miscellaneous support and subsidies
3.7%	Staff-operating expenditure – information
4.7%	Development co-operation
3.9%	Energy, industrial and transport research
4.2%	Regional policy
6%	Social policy
1%	Fisheries
4.1%	EAGGF Guidance and various agricultural funds

EAGGF Guarantee 65.3%

Total appropriations for commitments in 1978

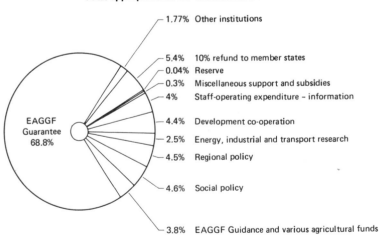

1.77%	Other institutions
5.4%	10% refund to member states
0.04%	Reserve
0.3%	Miscellaneous support and subsidies
4%	Staff-operating expenditure – information
4.4%	Development co-operation
2.5%	Energy, industrial and transport research
4.5%	Regional policy
4.6%	Social policy
3.8%	EAGGF Guidance and various agricultural funds

EAGGF Guarantee 68.8%

Source: Preliminary Draft General Budget of the EC for the Financial Year 1979, p. 70, **Diagram 4.**

Table 2.2   The Main Financial Instruments of the European Communities

	Legal Source	Date of Creation	Initial Revenue	Current Revenue	Objectives	Character
European Coal and Steel Community	Articles 49–56 (ECSC)	1951	Levies and borrowing	No change	Investment programmes, research, creation of employment, resettlement and retraining of workers	Grants, loans and interest subsidies
European Agricultural Guidance and Guarantee Fund	Article 40 (EEC)	1962	National contributions	Own resources	Financing of CAP price policy and special measures	Automatic intervention; total EC finance
Guarantee Guidance	Article 40 (EEC)	1962	National contributions	Own resources	Improve agricultural structures	Grants with national input to defined categories of farmers
European Social Fund Old Fund	Articles 123–8 (EEC)	1958	National contributions	—	Employment measures including resettlement and vocational training	Grants with national input to public and private agencies
European Social Fund Revised Fund	Articles 123–8 (EEC)	1971	Own resources	Own resources	Employment aids linked to EC policies and to particular categories, e.g. in problem regions	Grants with national input to public and private agencies
Regional Development Fund	Article 235 (EEC)	1975 (revised 1978)	Own resources	Own resources	Assist regional development by industrial and infrastructure measures	Grants with national input to public and private agencies; national quotas (except 5 %)

	Legal Source	Date of Creation	Initial Revenue	Current Revenue	Objectives	Character
Euratom	Article 174 (Euratom)	1958	National contributions and borrowing	Own resources and borrowing	Promotion of joint research and funding of Supply Agency	Direct projects and joint programmes
European Development Fund	Article 131 (EEC) and Implementing Convention	1958 (revised by Yaoundé and Lomé)	National contributions	National contributions	To promote economic and social well-being in countries associated under Article 131	Investment programmes and (Lomé) stabilisation of export earnings
European Investment Bank	Articles 129–30 (EEC) and Protocol on EIB	1958	National subscriptions and borrowing	No change	Regional modernisation and conversion measures; projects of common interest	Loans inside and outside EC
Ortoli Facility	Article 235 (EEC)	1979	Borrowing by Commission	No change	Promotion of energy industry and infrastructure	Loans for investment projects
European Monetary Co-operation Fund	Article 108 (EEC)	1971	Central bank pledges	No change	Reduce margins of fluctuations among national currencies	Loans to support currencies of member states; short term
Credit Facilities	Article 108 (EEC)	1970, 1971, 1976	Contributions from central banks and Commission borrowing	No change	Stabilisation measures for exchange rate and balance-of-payments problems	Credits and loans over very short, short and medium term

EC intervention, and then detailed negotiation of its constituents. Moreover there has been a shift of objectives from the early years, when it was thought appropriate to transfer the expenditure responsibility for particular policies from the national to the Community level, as was done for the maintenance of common agricultural prices. Recently the trend has been to introduce a parallel Community role to operate alongside national funding of the same or related activities. Table 2.2 summarises the chief features of the main financial instruments of the EC.

THE UNDERLYING ISSUES

It is important, therefore, to disentangle some of the central issues that underlie the discussion of Community expenditure. First, there is the fundamental question of what sorts of expenditures are logically required to underpin the establishment of a common market. Two different responses have figured in Community discussion. One school of thought takes the restrictive view that EC finances should be concentrated simply on those activities or sectors that are directly implicated. These would include, for example, cross-border projects, or modernisation schemes where a sector has suffered in an identifiable way from the removal of internal barriers. The more radical view holds that the common market has a far broader impact on economic activities and also diminishes the capacity of individual governments to intervene effectively, and that therefore fairly substantial Community involvement is required, for example, to redress regional imbalances.[9] In practice elements of both philosophies are evident in EC expenditure and indeed within particular programmes, including the RDF and the EIB.

A second issue is whether Community spending should be automatic or selective in its application. General programmes, such as the commitment to guarantee the prices of defined agricultural products in particular circumstances, mean that finance must be provided automatically if the relevant circumstances apply. In practice this makes precise estimates of likely expenditure almost impossible, especially given the variables that affect agricultural production and trade.[10] Equally it provides little margin for manoeuvre for the redeployment of resources. By contrast selective programmes permit the use of discretionary criteria as to allocations which may also be contained within fixed expenditure ceilings. This broadly has been the approach adopted for the RDF and the ESF. This distinction between the general and the selective approaches is linked to the issue of whether the EC should aspire to develop its own policies, in the sense of a broad framework for action within a sector and based on a complex of specific components, as has happened in agriculture. If the EC move in this direction then the expenditure consequences are inevitably substantial. The selective approach may have the more modest aim of limited intervention and a light burden on resources. The corollary of this, however, is that it then becomes far less easy to discern the economic impact of Community funding or to justify its extension, particularly since selective programmes are often linked to long-term projects.

EXPENDITURE THROUGH NATIONAL AUTHORITIES

The other major dimension of EC expenditure is the interaction between Community and national levels, which raises in turn important issues that have become enmeshed in the debates about convergence and equity. How much each country gets out of the Community has become an issue of high political salience, since the discussion of transfers through the budget depends on some assessment of the financial benefits accruing to member states from Community activities. The orthodox Community doctrine as expounded by the Commission has historically been that EC expenditure cannot and should not be broken down into national components, since this neglects the fact that expenditure serves Community rather than national policies. Thus the national destination of particular amounts of cash is less important than a judgement on the policy itself. In any case for various technical reasons it is extremely difficult to categorise expenditure 'objectively' into national balance-sheets.[11] However, in practice governments are inevitably interested in assessing the pattern and volume of their receipts from EC funds, in order to determine their attitudes to any proposed changes or innovations in expenditure. These concerns are often reflected not just in the general discussions of net transfers that have become topical, but also in the questions of additionality, conditionality and quotas.

Additionality is the doctrine that EC expenditure which is deployed in parallel to national programmes should clearly supplement national resources rather than substitute for national expenditure; in other words, governments should not draw on Community cash so as to avoid paying themselves. It has proved very difficult to demonstrate conclusively that Community funds have in practice been deployed additionally. The issue was for a time central in the debate over the establishment of the RDF.[12] Since the recession and the consequent cut-backs in public expenditure by governments it has begun to recede in importance, as it has been argued by several governments that certain programmes would have been significantly reduced had extra Community money not been available. While this may have defused the immediate sensitivity of the question, it does not remove the underlying issue of whether it is appropriate for Community resources to be utilised to top up national budgets as far as particular sectors are concerned. Nor is the issue at all clear-cut, given that the bulk of the grants disbursed from the EC budget, notably from the RDF, ESF and the Guidance section of the EAGGF, depend directly on joint intervention by the EC and public authorities within the member states. The consequence in some cases, particularly in the context of the ESF, is that governments in order to accommodate Community criteria find themselves funding in part specific programmes that they might not otherwise have introduced or retained. The problem of additionality therefore has implications for national budgets as well as importance for the principles of how Community resources should be deployed, especially as regards the expenditures on non-reimbursable grants to national authorities and organisations. In theory, it is far less of a problem in respect of the loan facilities of the EC, and in any case the money lent by the EC ultimately returns through repayments and can be recycled.

Conditionality is another key issue. Crudely, the issue is whether 'he who pays the piper calls the tune', that is, the authority that disburses finance should have a say in how the money is spent. With the general instruments of the EC such as EAGGF (Guarantee section) there is no problem, since money is spent only on defined activities. However, the provision of grants and loans cannot avoid the issue, since these financial instruments are established to pursue EC criteria, but are actually deployed by the other agencies to which EC institutions channel resources. Thus, for instance, the grants available from the ESF and RDF are based on legislative rules about their allocation, in order to preserve a degree of Community control. Again in practice this control is not easily achieved. The economic significance of this is that it is difficult to discern an identifiable economic impact from EC instruments, unless it can be guaranteed that resources will be used strictly for the objectives stated at the Community level. Furthermore it is difficult to justify increases in Community expenditure unless the record of existing programmes indicates that the financing is producing the intended effects, rather than being dispersed within the economies of the member states.[13] Conditionality is a fundamental principle as far as loans are concerned, and in practice is more easily applied to loans than to grants. A second application for a particular kind of loan can relatively easily be rejected, if the conditions applied to a previous loan have not been observed. However, it is important to note that the kind of conditions operated for EC loans have been narrow in scope, in that they concentrate primarily on criteria of financial viability rather than on policy objectives. Relatively few of the grant facilities of the EC are so specific that conditions can easily be applied with rigour.

Lastly there is the question of whether there should be any ratios to govern the distribution of particular EC funds to the member states in the form of quotas. The orthodox Community view has been that national quotas are inimical to the principle of EC financing, as, for example, are regional quotas in many national budgets. Quotas rob the financial authority at the centre of the opportunity to discriminate in the allocation of resources or to alter the pattern of their distribution flexibly from year to year or from programme to programme. Equally, quotas build in entitlements to a certain share of resources, and this renders more complex the application of principles of additionality and conditionality.[14] However, from the point of view of the recipient an explicit quota entitlement provides greater security of funding over a foreseeable period than a more discretionary system. Also, quotas may permit a greater decentralisation in the allocation of resources to specific programmes that allows a greater measure of flexibility and responsiveness to changing needs. The argument is thus finely balanced.

Within the current pattern of EC expenditure both of the propositions have been applied. Thus, for example, allocations from the ESF are based on policy priorities set at the Community level and the distribution of resources depends on the correspondence between national claims and Community criteria. By contrast the RDF is distributed primarily in the form of fixed national quotas, leaving room for manoeuvre to member governments within broad Community guidelines. The loan facilities do not carry national quotas and this has permitted wide variations in their national distribution in the case of EIB loans. Table 2.3 indicates the consequences of these arrangements. In addition it must be recorded that there is a scattering of selective EC instruments to which only some member states have access. These include the special oil facility, the new package of support for Mediterranean agriculture and, the recent interest rebate scheme for the 'less prosperous' participants in the EMS. Finally, the whole issue of quotas has become caught up in the debate about convergence and equity. The fixing of national quotas may be based either on proportionality of receipts or on the entrenching of redistributive criteria. So the decision to adopt quotas for the RDF permitted a concentration of resources in those member states with large regional disparities, whereas in the case of the ESF any redistributive character can be a function only of the way the criteria for resource allocation are determined. In attempting to ensure equity on both the expenditure and the income side of the budget some governments have seen the use of quotas as an important guarantee of access to Community resources.

Community Revenue

The pattern of revenue-raising in the EC has altered since their creation both quantitatively and qualitatively. Initially member governments committed themselves to make fairly modest national contributions to a small Community exchequer. The extension of policy agreements over the years has both increased the calls on revenue recognised at the outset and added a variety of new claims. Furthermore the EC has moved on from its reliance for revenue on national exchequers to acquire its 'own resources' based in effect on direct Community taxation on the economic activities of the member states. More recently it has become clear that the current revenue base will by 1982 be inadequate to support current programmes, let alone new programmes or comparable financial activity in new member states. A major issue

Table 2.3   Community Expenditure in the Member States

	EIB loans 1973–8 UA (m.)	EAGGF payments 1978 EUA (m.)		ESF Commitments 1978 EUA (m.)	RDF payments 1978 EUA (m.)	Estimates of expenditure 1980 EUA (m.)
		Guarantee	Guidance			
Belgium	107·0	601·8	15·7	11·1	6·15	1,460
Denmark	192·2	806·3	16·0	14·2	5·58	651
Germany	485·4	2,489·0	125·1	10·1	49·26	3,471
France	1,238·1	1,739·4	60·5	86·2	82·63	2,917
Ireland	361·2	552·2	16·8	44·4	33·15	573
Italy	2,472·0	747·2	31·3	233·1	220·33	2,621
Luxembourg	—	24·0	1·4	0·2	0·50	312
Netherlands	62·3	1,273·9	16·3	9·8	8·18	1,629
United Kingdom	1,889·1	439·0	40·5	111·2	148·59	1,561

Source: Annual reports of the EIB and individual funds; 1980 estimates derived from PDB 1980.

Note:

This table includes allocations only from the major sources of EC finance and therefore does not represent total allocations to member states. It records simply cash transfers according to the definitions of the Financial Regulation for accounting purposes. Thus the EAGGF figures include MCA payments on an 'exporter pays' basis. 1980 estimates include all expenditure from General Budget except a small part of operating costs. Belgian and Luxembourg figures inflated by administrative costs of Community institutions.

currently facing the EC is, therefore, whether a larger revenue base will be accepted by member governments, and if so to what sources of income it might be extended. The whole issue is further complicated by the question of how far EC revenue can or should be based on independent and objective criteria without taking into account the differential impact on individual member states. The orthodox Community view has been to stress objective criteria in order to establish the common fiscal property of the EC while in practice implicitly subjective criteria of what is acceptable as EC revenue determine the attitudes of member governments.

THE ORIGINAL SOURCES OF REVENUE

Initially the three separate Communities must be distinguished, since each had its own budgetary arrangements. It was not until 1968, as a consequence of the Merger Treaty, Article 20, that the EEC budget was combined with the operating budget of Euratom and the administrative budget of the ECSC into a General Budget.[15] The 1970 Treaty of Luxembourg brought in the research and investment budget of Euratom, but the rest of the operations of the ECSC remain separate. Similarly the EDF has been separately financed as has the Euratom Supply Agency, while the other loan and credit facilities have from the outset depended on special arrangements. The High Authority of the ECSC was given the right under the Treaty of Paris, Articles 49–50, to raise levies on coal and steel production up to the value of 1 per cent of total production and within prescribed constraints to raise further capital through loans. By contrast the Treaties of Rome in the first instance made the EEC and Euratom dependent on national contributions, and envisaged direct revenue-raising only at a later stage (Article 201, EEC; Article 173, Euratom). Table 2.4 shows the various keys to determine these contributions agreed in the original negotiations among the Six and in subsequent bargaining. The interesting point about these keys is that they incorporate both a recognition of differential economic capacities in the member states and a sensitivity to the political demands of the participating governments. Thus, for instance, Italian contributions to each of the programmes were relatively low except the operating budgets of the EEC and Euratom. But there were significant differences of emphasis for other member states too reflecting, for instance, different degrees of national involvement in the development of

Table 2.4   *Keys to Determine Gross National Contributions:*
Community of Six (%)

	Belgium	Germany	France	Italy	Luxem- bourg	Nether- lands
EEC and Euratom Operating budgets	7·90	28·00	28·00	28·00	0·20	7·90
ESF	8·80	32·00	32·00	20·00	0·20	7·00
Euratom Research	9·90	30·00	30·00	23·00	0·20	6·90
EAGGF 1962–5	7·95	31·67	32·58	18·00	0·22	9·58
EAGGF (1965–6)	8·25	31·70	28·00	21·50	0·20	10·35
General Budget (from 1970)	6·80	32·90	32·60	20·20	0·20	7·30
EDF I	12·04	34·41	34·41	6·88	0·22	12·04
EDF II	9·45	33·77	33·77	13·70	0·27	9·04
EDF III	8·89	33·16	33·16	15·62	0·28	8·89

Sources: Treaty of Rome (EEC), Article 200; Treaty of Rome (Euratom), Article 172; Decision of 21 April 1970, Article 3; and for EDF, Treaty of Rome, Implementing Convention on Part IV Association, and 1st and 2nd Yaoundé Conventions.

atomic energy or in overseas aid.[16] As spending programmes were revised or added, so the ratios of national contributions were moderated. The consequence was to produce a rather untidy pattern of revenue, but one which built in the subjective judgements of member states as to what constituted a 'fair' contribution.

The first attempt to move beyond this came in the mid-1960s, when in 1964/5 the Commission of the EEC drafted a package of proposals for financing the common agricultural policy over the long term. In its Memorandum of March 1965 the Commission proposed *inter alia*, in pursuance of Article 201 of the Treaty of Rome, that the EEC should acquire direct revenue.[17] The objective in budgetary terms was twofold: to gain financial independence from governments and to introduce fiscal instruments that would allow some flexibility of application, particularly since direct revenue would bear more heavily on some member states than others. The plan was therefore that customs duties and agricultural levies should become direct EC revenue, but, according to Commission estimates, this would generate more income than anticipated expenditure demands. The surplus would be redistributed to member states on the basis of economic policy judgements 'taking account of the economic and social situation in the various regions of the Community and of the need to ensure that burdens are *equitably* [author's italics] shared within the Community'.[18] The philosophy behind this was a reflection of national practice, and would have endowed the EC with the potential to deploy resources in a discriminating fashion that would cater to the needs of what in Community jargon were different regions, but in practice were different member states.

The French government objected strongly to this and to the other elements of the package. After the tense period of six months during which the French government boycotted the EC institutions, the Luxembourg compromise was reached on the political principle that the French maintained their right to veto agreements that threatened their total national interests. The outcome as far as EC finances were concerned reflected these tensions in that a much narrower financial arrangement was agreed based on direct national contributions, with the possibility of supplementing the income from levies on agricultural imports, but without any income surplus to expenditure requirements.[19] In effect the Council of Ministers, on French insistence, ensured that this would not permit any scope for balancing out the burden on individual governments except in so far as this was achieved by the key fixed for national contributions. Thus in practice the arrangements entrenched, at least implicitly, the concept of *juste retour* that the Commission had tried so hard to avoid. Quite apart from this major issue, the eventual decision had other unsatisfactory elements, given that Article 201 (EEC) had not been entirely respected and that the precise use of these resources in the EAGGF was a subject of dispute.

OWN RESOURCES

At the start the ECSC had access to direct revenue through the raising of levies on coal and steel production. In practice this has been set at a fairly modest level (0·29 per cent since 1972) which has not so far broken the 1 per cent limit prescribed in the Treaty of Paris. This, together with loan facilities and the interest from investments, comprises the working capital of its operating budget. After a period of significant activity during the 1950s the pace of ECSC financing slowed down during the 1960s. But as the recession of the mid-1970s began to bite into economic performance in the EC, so demand for involvement in both the steel and coal industries began to increase. Consequently the ECSC budget came under considerable strain as projected expenditure began to outstrip revenue. Nor did it seem appropriate in a period of economic constraint to increase the levy burden on industries such as steel that were undergoing great strain. The Commission, therefore, asked member governments to make up the operating deficit in 1978 through a special endowment, and proposed for the 1979 budget to add to ECSC revenue the customs duties that were levied on imports of coal and steel. It is important to note that the system of funding ECSC operations was based on complete independence from governments, even to the extent that the production levies are collected directly by the ECSC rather than through the agency of national administrations. This reflects the generally greater autonomy of the ECSC by comparison with the EEC and Euratom.[20] These provisions also reflect a general principle that Community revenue should be generated from economic activities appropriate to the sector in which they will be redeployed. Article 173 of the Euratom Treaty did envisage a similar system of levies, presumably to be based on atomic energy production, but the provision simply suggests that national contributions 'may be replaced, wholly or partly, by levies collected in the member states'. Consequently, Euratom has had to rely for its funds on national contributions supplemented by loans to finance research and investment (Article 172), and more recently on the revenue of the General Budget. The problems of financing Euratom have been compounded by protracted arguments over the funding of its research programme.

Towards the end of the 1960s an accumulation of pressures reopened the question of raising revenue directly from independent sources for the General Budget. The completion of the customs union in 1968 and the end of the transitional period of the EEC in 1970 demanded a decision on the treatment of customs duties. A concern on the part of several governments to extend expenditure significantly beyond the agricultural sector, notably towards an economic union, increased interest in expanding the financial capacity of the EC. The resignation of General de Gaulle in April 1969 led to a modification of French policy on this as on other issues. In July 1969 the Commission produced a formal proposal to replace national contributions once and for all by the progressive introduction of a full system of 'own resources'.[21] Two sources of direct income were obvious: customs duties, following Article 201 (EEC), and agricultural levies, in the light of the precedent set by the temporary arrangements for the EAGGF. It was quite clear, however, that these alone would provide a highly unsatisfactory revenue base. In principle both levies and duties would decline, as the EC came closer to agricultural self-sufficiency and as world tariff negotiations brought down duties. Nor does either reflect directly changes in economic output. Certainly they gave little scope for the development of a larger budget, even if, as some suggested, the level of expenditure on agriculture would decline in real terms. An additional source of revenue, ideally with some elasticity and some relation to economic growth, was necessary. At first the Commission did not commit itself, though several possibilities were floated including fuel and pollution taxes. Eventually the Dutch suggested that a proportion of the national receipts from Value Added Tax (VAT) become EC property. The merits of VAT were, first, that all member states were already committed to its introduction in a largely similar form and, secondly, that its yield would increase with economic growth. The French were insistent that a firm agreement should be reached which would not allow the European Parliament a voice on which sources of revenue should go to the EC. Consequently the Hague Summit of December 1969 endorsed in outline the proposals for 'own resources' derived from these three sources – levies (or other charges arising from the CAP), duties and VAT – and the final version of the agreement was reached in April 1970.[22] Other taxes that arose as a consequence of other common policies might also contribute eventually to the 'own resources'.

The decision thus freed the EC from dependence on national exchequers, by establishing the principle of a Community right to call on its own revenue sources according to independent and objective criteria. It is, however, important to stress the difference in character between duties and levies on the one hand and the access to a share of national VAT receipts. Duties and levies are a direct fiscal consequence of Community procedures. Agreements on the Common External Tariff and on the CAP have included as an integral element the imposition of duties and levies at the external boundary of the EC, irrespective of the point of entry or the destruction of the imports to particular member states. National governments enter in only in so far as they act as the agents of the EC in collecting the revenue. Even though geographical factors might in practice mean that a particularly large share of receipts was collected in an individual member state, notably among the Six in the Netherlands, this was not to detract from their categorisation as collective Community property. The so-called Rotterdam principle applies, based on the recognition that though Rotterdam acts as important point of entry for goods in transit to member states other than the Netherlands the levies and duties collected there are not the national property of the Netherlands nor a particular burden on the Dutch economy.

It is not, however, possible to apply this doctrine in a pure form to the VAT element of 'own resources', since VAT remains a national rather than Community tax, though subject to

Table 2.5   *Keys to Determine Gross National Contributions:*
Community of Nine (%)

	1973/4	*1975*	*1976*	*1977*	*1978*[2]	EDF
Belgium	5·28	4·11	4·07	4·14	4·49	6·25
Denmark[1]	2·46	2·53	2·52	2·40	2·58	2·97
Germany	25·53	29·87	29·58	28·60	32·13	25·95
France	25·30	22·99	23·05	23·29	23·88	25·95
Ireland[1]	0·61	0·67	0·70	0·71	0·60	0·60
Italy	15·60	14·77	14·80	15·21	13·20	12·00
Luxembourg	0·16	0·16	0·15	0·17	0·17	0·20
Netherlands	5·66	5·26	5·32	5·32	6·00	7·95
United Kingdom[1]	19·32	19·64	19·80	20·04	16·90	18·70

Source: For EC General Budget, based on Treaty of Accession, Article 129, which amended the 'own resources' decision of 1970; for the EDF, based on the Lomé Convention.

Notes:
[1] Article 131 of the Treaty of Accession staggered the proportions of the contributions to be paid by the new members, 1973–7, with an extra limitation for two further years, 1978–9.
[2] Originally 1978 was the year in which GNP-based contributions were to have been replaced by VAT contributions.

Community rules.[23] Governments thus retain the right within these rules to vary the levels of VAT charged and the goods to which they apply. But the establishment of these 'objective' criteria for determining EC revenue does not prevent significant discrepancies between gross national contributions to the EC budget, according to those three elements, and the relative weight and success of member states' economies. Thus the victory from a Community viewpoint of establishing the new system was double-edged: independence from national exchequers was entrenched, but at the cost of a loss of the balancing factors that had characterised the old system. This lack of flexibility carries a built-in danger of the EC not being able to adjust its tax burden to the different needs of their component parts. In practice the introduction of the full 'own resources' system was not to be achieved until 1980, when all member states would at last have introduced the harmonised base of assessment for VAT on which the Community precept was to be made up to a ceiling of 1 per cent.[24] Table 2.5 shows the interim pattern of national contributions that operated prior to full 'own resources', adjusted for the Community of nine members.

THE ACCESSION OF NEW MEMBERS

The seal was set on the new revenue system only months before accession negotiations opened between the EC and Denmark, Ireland, Norway and the UK. 'Own resources' thus comprised part of the *acquis communautaire* which the candidates were obliged to accept. It is simple in theory to argue that new members of a club should agree to accept the established rules. However, the realities of politics and economics may undermine the theory, if those rules seem to place a heavier relative burden on some than others. This was a source of concern in the UK at the outset, where it was feared that on the income side of the budget alone the UK government would eventually be making an extraordinarily high contribution. The estimates made in the UK reflected in particular the assumption that high levels of both agricultural and industrial imports from outside the EC would persist and that trade diversion to give preference to imports from other EC countries would occur only gradually.[25] The Commission's response and that of other member states was to argue Community preference for imports of both agricultural and industrial goods, thus diminishing the burden of levies and duties. In addition they assumed that the revenue contributed to the budget from VAT receipts would grow as a proportion and in total volume, and that this would bear on the UK economy at comparable levels to its impact on other member states. Eventually a formula was agreed in the Act of Accession of 1972, Articles 127–32, whereby the contributions of all new members would be staggered over the five years of the transitional period, with a further limitation in Article 131 on the rate of increase for two more years.[26] A further caveat was conceded to the British informally that if 'unacceptable situations' should arise, 'the very survival of the Community would demand that the institutions find *equitable* solutions'.[27] A similar provision has been included in the Treaty of Accession with Greece of May 1979. The principle of such an arrangement is to allow time for the expenditure policies of the EC to adapt to the needs of new members. Its effect, however, may simply be to mask the impact of the contributions from new member states until the end of the period of transition, so that neither the economic nor the political repercussions are immediately evident.

However, in the British case the provisions of the Treaty of Accession were soon disputed with the arrival in office in 1974 of a Labour government, committed to renegotiate the terms of EC membership. The estimates of the high budgetary burden on the UK had been a particular preoccupation within the Labour Party, and thus became a major plank in the renegotiation platform. Eventually at the Dublin session of the European Council in March 1975 a complex formula was agreed whereby a new Financial Mechanism was adopted that would allow a corrective repayment in respect of unduly heavy gross contributions.[28] This sought to meet the British case, which rested on the projected disproportion between UK gross contributions and the UK's share of Community GNP. The mandate given to the Commission for devising the formula made reference to the need for avoidance 'during the process of convergence of the economies of member states of situations unacceptable to a member state and incompatible with the sound functioning of the Community'.[29] Thus the British argument was explicitly linked not just to equity but to both convergence and the general health of the system.

The mechanism itself is extremely complicated and contains three important pre-conditions to determine 'an unacceptable situation' and intricate rules about the consequent calculations. Three points must be stressed. First, the mechanism was of general application and not confined to the UK. Some commentators hailed it as an important breakthrough in the sense of incorporating equity linked to the GNP shares into the structure of budget revenue. Secondly, it applies solely to gross contributions, though the net position of member states was included as an element in the calculation of entitlement to benefit from the pay-back mechanism. This, in itself, was a major departure from Community orthodoxy which had hitherto prevented any allusion in EC legislation to even a hint of *juste retour*. Thirdly, it restated the old doctrine that levies and duties were automatically collective property, since the repayment could not exceed the member states' VAT contribution to the budget, even if the distortion by comparison with GNP was actually generated by payments of levies and duties. In practice the caveats that apply to the mechanism prevented it from being activated before 1980 in respect of the 1979 budget. Changes in economic circumstances had meant that the criteria for its application designed with the UK problem in mind did not in practice allow it to operate earlier, and even then the sums of money involved were relatively trivial.[30] Table 2.6 shows the breakdown of gross national contributions to the 1979 budget according to one of the Commission's estimates. The EC Committee of the House of Lords recommended in June 1979 that a revised form of the Financial Mechanism would be necessary to redress the balance of UK contributions to the EC budget. In July 1979 the British government began publicly to argue the same case with its Community partners, but making it clear that any new Financial Mechanism would need to take account of net national contributions to the budget.

FUTURE SOURCES OF REVENUE

Even as the initial 'own resources' system was being set in place it became clear that they would soon cease to be adequate to support existing expenditure commitments based on current policies. In its document 'The Way Ahead' published in November 1978, the Commission stated that on most realistic assumptions the ceiling of 1 per cent of VAT contributions to the budget would be reached in 1981, even without taking into account the needs of new

Table 2.6 The Financing of the 1979 Community Budget

	Agricultural Levies (including sugar and isoglucose) EUA (m.)	%	Customs Duties EUA (m.)	%	VAT Contributions[1] EUA (m.)	%	Total EUA (m.)	%
Belgium	242·6	11·16	318·7	6·72	311·0	4·68	872·3	6·43
Denmark	46·5	2·14	118·9	2·51	174·9[2]	2·63	340·3	2·51
Germany	469·6	21·61	1445·8	30·47	2116·0[3]	31·87	4031·4	29·73
France	240·0	11·04	718·1	15·13	1622·3	24·43	2580·4	19·03
Ireland	14·7	0·68	50·0	1·05	39·3[2,3]	0·59	104·0	0·77
Italy	410·2	18·88	451·3	9·51	706·9	10·65	1568·4	11·57
Luxembourg	0·1	0·005	3·9	0·08	14·1[3]	0·21	18·1	0·13
Netherlands	371·7	17·11	448·7	9·45	427·9	6·44	1248·3	9·21
United Kingdom	377·6	17·38	1190·1	25·08	1227·09[2]	18·49	2795·6	20·62
TOTAL (including % shares of each source of revenue)	2173·0	16·03	4745·5	35·0	6640·3	48·97	13558·8	100·00
Miscellaneous Revenue							143·9	
Total Budget							13702·7 [£8,750m.]	

Source: General Budget of the European Communities for 1979, including Supplementary and Amending Budgets Nos 1 and 2 (percentages added).

Notes:
1 This represents a precept of 0·744 per cent of the VAT yield on the harmonised base of assessment.
2 These contributions are further subject to the provisions of Article 131 of the Treaty of Accession.
3 Financial contributions in lieu of VAT.

member states.[31] In practice VAT had not proved sufficiently buoyant to allow much room for increased expenditure programmes, particularly given the fiscal consequences of reduced economic growth. Furthermore the level of agricultural expenditure and its proportion of total expenditure had remained higher than anticipated. This left open three options: an irrevocable predominance of agricultural expenditure; positive reductions in agricultural expenditure to allow more expenditure on other policies; and thirdly the addition of extra revenue into the budget. The Commission clearly favoured the last, though it simultaneously recognised a need to keep agricultural spending within bounds. Three methods of finding extra finance were identified: first, more extended borrowing, though this is circumscribed by the Treaty prohibition of deficit financing for the general budget; secondly, the reintroduction of direct national contributions either as a supplement or to fund any new policies; and thirdly, additional direct revenue with the character of 'own resources'. The Commission document then discussed various candidates for this latter course, concluding that in the short and medium term the EC 'will have to rely predominantly on VAT as a source of revenue'. It stated quite categorically its dislike of national contributions, precisely because of the political difficulties that this would provoke.

The document also addressed itself to the twin problems of convergence and equity. It argued that Community revenue-raising should take into account the broad principle of taxable capacity or ability to pay in that the burden of taxation should be equitably distributed. To this end various indicators were recorded that might offer criteria of ability to pay in *proportional* terms (see Table 2.7). However, the more radical option of establishing a *progressive* fiscal regime would have to be linked to the objective of seeking positively to reduce economic disparities. This in turn would require some adjustments to both the income and the expenditure sides of the budget. In respect of income the Commission first set aside duties and levies as not relevant to the determination of taxable capacity, because they comprise 'own resources' in a pure sense. But the paper went on to recognise the regressivity of VAT in some cases, and thence to suggest that a corrective mechanism might be applied that would offset any undue burden on some member states, a device that has been dubbed 'the progressive key'. A proposal on these lines would

Table 2.7 Criteria of Ability to Pay

Relative sizes of member states' income and consumption in 1976

	GDP	National income	Private consumption	Personal income	Per capita GDP as % of EC average	Population 1,000	%
	as % of EC total						
Germany	32·14	31·70	31·78	31·40	135	61,531	23·78
France	25·03	25·00	25·37	23·72	122	52,921	20·45
Italy	12·24	12·50	13·22	12·30	56	56,169	21·70
Netherlands	6·44	6·57	6·95	6·74	121	13,773	5·32
Belgium	4·75	4·88	3·70	4·23	125	9,818	3·79
Luxembourg	0·17	0·15	0·15	0·18	123	356	0·14
United Kingdom	15·91	15·77	15·58	17·98	73	56,001	21·64
Ireland	0·57	0·62	0·60	0·57	47	3,162	1·22
Denmark	2·75	2·81	2·65	2·87	141	5,083	1·96
EC 9	100·00	100·00	100·00	100·00	100·00	258,805	100·00

Source: SOEC National Accounts 1977, reprinted in *The Way Ahead*, *Bulletin of the European Communities* Supplement 8, 1978, p. 17, Table 6.

entrench a formula that would follow in part the aims of the 1975 Financial Mechanism but apply them in a rather different way. Thus implicitly the British concern was voiced, though with the recognition by the Commission that there was not as yet any overall consensus shared by all member governments that such a move would be acceptable.

Discussions are proceeding only very slowly on these suggestions. Strong differences of view persist among the member states over the principles involved. These are in turn affected by other considerations. One issue is whether the problem of the discrepancy between revenue shares and ability to pay should be viewed as a Community problem rather than a difficulty of particular member states. A second issue is whether it is in any case desirable for the overall size of the Community budget to increase. There is strong evidence, for instance, that the preference of the West German government is to resist any expansion beyond the resource base already agreed.[32] This in turn is linked to concerns that are not restricted to West Germany that public expenditure in general should be held within bounds. Nor does there seem to be much indication that a Community-wide consensus is yet in prospect that would permit a significantly larger budget to emerge at the EC level designed to replace portions of national budgets, rather than supplement national programmes.

THE FINANCING OF OTHER INSTRUMENTS

When the Spaak Committee was drafting the provisions of the EEC Treaty, the negotiators gave some consideration to the establishment of a regional fund to work on parallel lines to the Social Fund. Instead they opted for an 'Investment Fund', whose task would be 'to participate in the financing of projects which, by their scope or simply by their nature, do not lend themselves easily to the various methods of financing available in each individual state'.[33] This became transmuted into the European Investment Bank, an independent financial institution operating under a separate protocol to raise capital that could be lent on to serve various defined EC objectives.[34] This independence is reflected in the capital base of the EIB which was given an initial subscription of 1,000 million UA, of which the first 250 million was paid in by member governments. The ratios contributed by each government were 30 per cent each from Germany and France, 24 per cent from Italy, 8·65 per cent from Belgium, 7·15 per cent from Netherlands and 0·2 per cent from Luxemburg.

During the 1960s the EIB began to borrow further capital on the markets, helped by its winning on the New York market a coveted 'triple A' rating for creditworthiness. In 1971 the subscribed capital rose to 1,500 million UA, increased to 2,025 million by the 1973 enlargement, to 3,543·75 million in 1975, and to 7,087·5 million in June 1978. The subscription ratios for the nine are Germany, France and the UK, 22·2 per cent; Italy, 17·7 per cent; Belgium and the Netherlands, 5·85 per cent; Denmark, 2·96 per cent; Ireland, 0·74 per cent, and Luxembourg, 0·15 per cent. Currently only 15·7 per cent of the total subscription notionally demanded from member states is actually called in by the EIB. These substantial increases in the EIB's activities during the 1970s have arisen from vastly expanded claims on its resources, accelerated by an attempt on the part of EC institutions collectively to use the EIB's facilities to reinforce other Community objectives. This contrasts strongly with the EIB's earlier image as a rather conservative lender, though it retains considerable caution in order to preserve its high credit status. In practice, however, a significant number of EIB loans have been relatively 'soft' in terms of their conditions and use, by comparison with private credit institutions. It must, however, also be stressed that member governments have on the whole more easily accepted sizeable increases in EIB capital than extensions of revenue for the Community budget. This is partly because in practice governments are not regularly required to pay large sums of cash to the EIB, and partly because the cash is recycled back through the EIB to be made available for further projects on a continuing basis.

These factors help to explain the reasoning that led to the recent increase in EC borrowing and in particular the creation of the new loan system, 'the Ortoli facility'. The ECSC had under its Treaty the right to raise loans, a facility that was used only marginally until 1972 when a rapid increase occurred. Euratom also had the right to borrow, but the facility has not been exploited significantly.[35] In response to the oil crisis of 1973 a further facility was developed enabling the EEC to raise capital to help member states to cover the short-term costs of increased energy prices.[36] In early 1977 the European Council asked the Commission and EIB to consider ways in which the various financial instruments of the EC could be made more effective in combating unemployment, increasing investment and promoting economic convergence. After protracted negotiations the new facility was established in May 1979 as a joint EC/EIB instrument.[37] This empowers the Commission to raise 1,000 million European units of account in loans to be managed by the EIB in accordance with criteria determined by the Council of Ministers. The assumptions behind the proposal were first that there were certain types of important project, particularly of an infrastructure character, which could not easily find investment capital from existing sources, but secondly that an EC mechanism might be able to tap the financial markets and then lend on to appropriate projects. The philosophy thus is much the same as that which underlay the EIB's creation, with the difference that in the late 1950s financial markets were buoyant, though in danger of neglecting particular kinds of investment, whereas by the mid-1970s economic uncertainties might deter such investment, unless positive incentives were offered. From this came an acceptance among member governments that a new hybrid facility which brought into play the different EC institutions would permit an experiment in deploying additional financial resources. It must, however, be noted that the decision to make the facility a loan operation rather than a new fund represents a degree of disenchantment with the more orthodox budgetary operations of the EC. An underlying issue for the analysis of the Ortoli facility is whether it is really a new departure for the EC in terms of its scope and objectives, or whether it is rather a cosmetic response to the economic problems of the mid-1970s based on the limited model of the EIB.

TECHNICAL ISSUES

The analysis of the EC budget is made more intricate by a number of complex technical problems. Perhaps the most acute is the difficulty of forecasting with precision the level of projected expenditure from the General Budget. The chief problem arises from the heavy burden of agricultural expenditure on price guarantees, the costs of which are affected by a range of factors outside the control of the EC budget. Since the revenue side of the budget is deduced from expenditure estimates, this makes the whole process both complex and subject to error. If expenditure rises during the financial year further revenue has to be called in. This may be derived from reserves or transferred from other parts of the budget. However, it has been common for supplementary budgets and amending letters to be drafted by the Commission as and when their estimates of agricultural expenditure become firmer. Under the old system which included national contributions member governments could be asked to produce a further payment. Under the full 'own resources' system a further precept on VAT can be called in from member states but not, so far, beyond the 1 per cent ceiling. This new constraint will therefore put an extra premium on reliable forecasts of expenditure, particularly given the pressures to direct finances to other sectors.

A further problem is the division of expenditure into *compulsory* and *non-compulsory* categories, compulsory being the expenditure that necessarily arises from commitments in the Treaties, and non-compulsory comprising, as it were, the more optional elements in the budget.[38] This was established in the 1970 Treaty. In practice the importance of this is political and procedural rather than economic, and relates primarily to the power of the Parliament to amend appropriations. The distinction roughly correlates with the difference between automatic policy support and the more selective and the narrower financial instruments. Compulsory expenditure now includes primarily price support under the CAP and certain kinds of foreign aid under EC treaty obligations to third countries. Each new area of spending has to be classified as either compulsory or non-compulsory, in order to determine which procedures govern its consideration by EC institutions.

Another issue is the decision taken each year on the overall increase in the total size of the General Budget. The provisions of the Amending Treaty of 1975 have a requirement that the *maximum rate* of increase for *non-compulsory* expenditure should be set according to several criteria, agreed with the budgetary authorities, and then used to govern planned expenditure.[39] Each June the Commission declares the *maximum rate* applicable to the next year's budget. Though this provision is related to economic criteria, it has in practice been more interesting as a source of political differences. The *maximum rate* may be explicitly increased by majority decisions of both the Council and the Parliament. In addition the Parliament has a further *margin of manoeuvre* to increase the total of *non-compulsory* expenditure within certain limits, beyond the increases put forward by the Council. An extra complication is the introduction of a distinction between *payments* and *commitments* in the appropriations for each budget.[40] At one time the budget, except for Euratom, simply represented the precise payments to be made in each year, but the attempt to look at expenditure plans over several years led to the addition of a 'commitments' column to include the appropriations of EC finances to certain programmes looking several years ahead. Only 'payment' appropriations for each year need to be covered out of income. This has to be seen in conjunction with the broader attempts by the Commission to examine economic trends and their budgetary consequences in triennial forecasts and 'global appraisals' presented to the Council of Ministers.[41]

An extra set of complications are the direct consequences of changes in the international monetary system and the alteration in the exchange rates among the currencies of member states. Initially Community finances were denominated in notional units of account (UA) linked to gold and thus in practice to dollar values. By the late 1960s this ceased to be a straightforward matter and for agricultural guarantees a separate agricultural unit of account, still linked to gold but different from the budget UA, was introduced along with the system of Monetary Compensatory Amounts (MCAs) or 'green money' to compensate member states for the effect of exchange rate fluctuations. Two difficulties thus arise: first, the lack of unity in the values of different units of accounts for budgeting purposes and, secondly, the appropriate evaluation of MCAs in the expenditure side of the budget. Attempts have been made to restore unity for budgeting purposes with the creation of a new European Unit of Account based on a 'basket' of the member states' currencies appropriately weighted. This was devised in 1975 and extended to the General Budget in December 1977. It was extended to cover the bulk of financial instruments, though it did not fully incorporate the special agriculture UA. More recently the creation of the new EMS has resulted in the denomination of the EUA in the new European Currency Unit (ECU), based similarly on the EMS basket of national currencies, and this has taken in the expenditure on agricultural guarantees.[42]

These distinctions are important not just for technical reasons but because the effect of using different denominations can be to attribute markedly different values to particular categories of expenditure. Equally thorny is the question of how to show MCAs in the budget, partly because they have considerably inflated the volume of expenditure on agricultural price support, but mainly because there has been considerable controversy over the way in which they are attributed to different member states.[43] A variety of different formulae have been used to deal with MCAs and to determine the appropriate point to make the actual payments. Since 1976 MCAs have been paid or charged in the country the currency of which gives rise to the MCA. However, in the case of exports to the UK and Italy all MCAs are actually paid out in cash in the country from which the export is made, even though they may arise from the prevailing levels of the green lira and the green pound.

It has been argued by the Commission and some member governments that the effect of MCAs is to reduce consumer prices in importing member states, and thus their value should be counted as a benefit to these two importing countries, even though the cash transactions take place elsewhere. By contrast, the UK and Italy have argued that they are not budgetary benefits to them. The difficulty is that different attributions of MCAs significantly alter the net contributions to and receipts from the budget of individual member states. However, it is important here to distinguish the budgetary effects from the economic effects. MCAs as budget payments may actually be spent in the exporting member states, but they may none the less give some economic benefit to importing countries in so far as consumer prices and the prices paid for the goods across the exchanges are thus lower than they would be in the absence of MCAs. It is, however, important to note that the MCA problem diminishes if certain changes take place in the market exchange rates. Thus in the summer of 1979 the rise in the value of the pound sterling eliminated the use of green rates for the UK.

Aggregating Revenue and Expenditure

The pattern that has now emerged is of a base EC budget that reflects on the expenditure side a number of continuing programmes. Agricultural financing still consumes the largest share by far, though social and regional spending have now established a continuing claim on resources. Major new programmes cannot, however, be incorporated into the budget without an explicit alteration of the base itself. This requires not simply agreement that new programmes are desirable *per se*, but also an acceptance that it is legitimate to enlarge the revenue base itself. Increasingly the debate has shifted to focus on the implications of changes on both sides of the budget for the net contributions and receipts of individual member states. Though the Commission has resisted pressures to produce national balance-sheets, these have come in practice to figure more and more explicitly in negotiations over the composition of the budget. Tables 1.1(a) and 1.1(b) (p. 28) provide information on the relative positions of individual member states in recent years. The precise sums have to be treated with great caution, but the trends and relative shares are now fairly widely accepted. The controversy continues, however, over what political and economic conclusions should be drawn from them.

Notes: Chapter 2

1   Most of this literature is British, given the particular concern in the UK over the impact of net contributions to the EC budget on the economy. See the periodic assessments in the *National Institute Economic Review* of British economic trends including EC effects and, in particular, A. E. Daly, 'UK visible trade and the Common Market', *National Institute Economic Review*, November 1978, and the *Cambridge Economic Policy Review*, March 1978 and April 1979. Relatively little detailed work has been published by the Commission of the EC, though a Commission study on the regional effects of the CAP suggested that richer regions were benefiting considerably more from some programmes than poorer regions. See *The Economist*, 13 November 1976.

2   See below, pp. 106–9.

3   See, for example, H. Wallace, W. Wallace and C. Webb (eds), *Policy-Making in the European Communities*, Chichester, Wiley, 1977, ch. 3; and R. A. McAllister, 'Ends and means revisited: some conundra of the Fourth Medium-Term Economic Policy Programme', *Common Market Law Review*, vol. 16, no. 1, February 1979, pp. 61–76.

4   See Etienne-Sadi Kirschen *et al.*, *Financial Integration in Western Europe*, New York, Columbia University Press, 1969.

5   See *Twelfth General Report on the Activities of the European Communities*, Brussels, Commission, February 1979, pp. 115–18.

6   There are difficulties about this, because the EDF is still funded directly by national exchequer contributions committed over a five-year period. The arguments for its inclusion are that the EDF would then be subject to parliamentary control, and that overseas development aid would figure as a significant proposition of the total budget, thus decreasing the share of agriculture.

See Daniel Strasser, *The Finances of Europe*, New York, Praeger, 1977, pp. 94–5.

7   See below, p. 68, on the Unit of Account, its replacement by the European Unit of Account, and the link to the European Currency Unit.

8   For details see Strasser, op. cit.

9   The first school of thought tends to reflect the thinking of the governments of the 'prosperous' member states, while the second tends to be the view of the Commission and to some extent of the governments of the 'less prosperous' countries. For a discussion of the underlying issues see Michael Hodges (ed.), *Economic Divergence in the European Community* (provisional title), London, Allen & Unwin, forthcoming.

10 The effects of climate, harvest and changing acreage or animal stocks on agricultural expenditure cannot be forecast with precision, or much in advance of the actual disbursement of expenditure.

11 Problems arise on the expenditure side of the budget over the rate at which payments are actually made to fund EC policies in the member states, over the arbitrary allocation of certain expenses for purely Community programmes (such as the administrative costs of EC institutions in Belgium and Luxembourg or the food aid scheme) that are not really credits to member states, over the attribution of MCAs, and over the exchange rate values of the UA or EUA. See *Community Budget 1978 – Net Cash Transfers*, ISEC/B23/79, London, Commission Office, 24 May 1979.

12 The argument over additionality was mentioned in the Preamble to Regulation (EEC) No. 724/75 of the Council of 18 March 1975 establishing a European Regional Development Fund, *Official Journal* L73 of 21 March 1975, 'whereas the Fund's assistance should not lead Member States to reduce their own regional development efforts, but should complement these efforts'. See also annual reports on the ERDF.

13 Conditionality has been particularly stressed in German thinking on EC finances, where an analogy is drawn with the concept of *Mitspracherrecht* used domestically to require involvement by the provider of resources in determining policy to be implemented by another authority. The principle was actively extended to the EC level in the various directives since December 1974 on the provision of credits for the Italian economy from an EC facility, roughly along IMF lines. The French and British governments have been particularly reluctant to see this principle extended to EC expenditure across the board.

14 Where quotas are set there is a tendency for governments to submit projects for EC funding only up to the ceiling of their overall quotas, thus making it difficult for the EC authorities to choose priorities or to apply conditions systematically.

15 Treaty Establishing a Single Council and a Single Commission of the European Communities, signed in 1965 and operable from July 1967.

16 This is reflected in the different numerical weights used for the purpose of qualified majority votes. See below, Table 3.2, p. 85.

17 Proposal submitted by the Commission to the Council on 31 March 1965, 'Financing the Common Agricultural Policy – independent revenue for the Community – wider powers for the European Parliament', *Bull. EEC*, Supplement 5, 1965, pp. 2–11.

18 'Proposal for provisions to be adopted by the Council by virtue of article 201 ...', Article 5, *Bull. EEC*, Supplement 5, 1965, p. 10.

19 Council Regulation 130/11/EEC, 26 July 1966.

20 See Strasser, op. cit., pp. 116–18. Centralised management is also possible for the relatively small number of steel producers, but not for the intricacies of collecting the other sorts of EC revenue. On the financing problems of Euratom, see Wallace, Wallace and Webb, op. cit., p. 182.

21 *Proposal of the Commission for the Creation of Own Resources for the Communities*, Doc. 99, 1969/70-COM(69)700, 16 July 1969. For related discussions and texts, see *Les ressources propres aux Communautées européennes et les pouvoirs budgétaires du Parlement européen*, Luxembourg, European Parliament, June 1970. See also D. Coombes and I. Wiebecke, *The Power of the Purse in the European Communities*, London, RIIA/PEP, 1972, pp. 26–9.

22 See the Communiqué of the Hague Summit of 2 December 1969, para. 5, reprinted in *Les ressources propres ...*, p. 135; and the Decision of 21 April 1970 on the replacement of financial contributions from member states by the communities' own resources, *OJ* L94, 28 April 1970, p. 196.

23 For a discussion of VAT harmonisation see D. Puchala, 'Worm cans and worth taxes: fiscal harmonisation in the European policy process', in Wallace, Wallace and Webb, op. cit., pp. 249–72.

24 Agreement on the harmonised base was eventually reached on 17 May 1977 in the Sixth Directive on VAT, *OJ* L145, 13 June 1977. Delays in implementation prevented its application to the 1978 budget. For the 1979 budget the majority of member states made VAT contributions, but not Germany, Ireland and Luxembourg, the contributions of which were GNP-based. In July 1979 the Commission was considering taking the German government before the Court of Justice for its continuing failure to implement the Directive; Germany complied in November.

25 See, for example, the accounts in G. Denton *et al.*, *The Economics of Renegotiation*, London, Federal Trust, May 1975.

26 Article 130 set a limitation on contributions from the new member states in staggered proportions of their total liability (45 per cent, 1973; 56 per cent, 1974; 67·5 per cent, 1975; 79·5 per cent, 1976; and 92 per cent, 1977); Article 131 set an extra limitation for 1978 and 1979 to prevent too sharp an increase in the actual volume of payments by comparison with 1977. A further complication was the application of the new European Unit of Account, under a decision of 21 December 1977, to the General Budget. This raised in real terms the volume of the UK contribution, and a further formula to permit repayments was agreed in December 1977 by the Council. See the *Second Report of the Select Committee on the European Communities of the House of Lords*, Session 1979/80, p. xlvi.

27 The informal concession on equity is recorded in *The United Kingdom and the European Communities*. Cmnd 4715, London, HMSO, July 1971.

28 This was embodied in Council Regulation (EEC) No. 1172/76 of 17 May 1976 setting up a Financial Mechanism, *OJ* L131/7-8, 20 May 1976. For comments, see John Dodsworth, 'European community financing: an analysis of the Dublin Amendment', *Journal of Common Market Studies*, vol. XIII, no. 3, December 1975, pp. 129–139; and M. R. Emerson and T. W. K. Scott, 'The Financial Mechanism in the budget of the European Community: the hard core of the British "renegotiations" of 1974–75', *Common Market Law Review*, 14, 1977, pp. 209–29. For the British government's view, see *Membership of the European Communities: Report on Renegotiations*, Cmnd 6003, London, HMSO, March 1975, paras 33–45.

29 Communiqué of the Paris Summit of 11 December 1974, *Bull. EC*, 12/1974, p. 12; and the Commission's proposal, *The Unacceptable Situation and the Corrective Mechanism*, Doc. COM(75)40 final, 30 January 1975.

30 *Second Report ... of the House of Lords*, 1979/80, pp. xvi, xxx–xxxi and xlvii–xlviii.

31 'Financing the Community budget: The Way Ahead', *Bull. EC*, Supplement 8/78 (reprint of Commission Communication to the Council of 23 November 1978).

32 See, for example, 'Germans may favour farm policy reform', *Financial Times*, 15 March 1979, and 'Powerful pressures for EEC farm reform', *Financial Times*, 8 June 1979, which report the concern of Chancellor Schmidt and the German finance ministry to resist increases in EC revenue. There is sympathy for this position in France, and evidence that the Conservative government elected in May 1979 in the UK supports it too. The German concern goes back at least to the autumn of 1974, when the German government tried to slash the appropriations for the 1975 budget. Part of the explanation for this concern stems from the increasing size of the German national budget deficit, a significant proportion of which can be notionally blamed on heavy German contributions to the UK budget.

33 The Spaak Committee report, Brussels, 1956.

34 On the establishment of the EIB, see the Protocol on the Statute of the European Investment Bank attached to the EEC Treaty. See also European Investment Bank, *20 years 1958–1978*, Luxembourg, EIB, 1978. There is no thorough academic analysis published on the EIB.

35 See Strasser, op. cit., pp. 133–8.

37 'The Ortoli facility' or 'New Community Instrument' was finally approved in a Decision of 16 October 1978, *OJ* L298, 25 October 1978 (see *Bull. EC*, 10/1978, pp. 21–2), and the Council of Ministers gave approval on 14 May 1979 for the Commission to raise the first 500 million EUA on the capital market, to be used in the first instance for energy and infrastructure projects.

38 See Strasser, op. cit., pp. 33–4; below, pp. 76–7, 83–4; and *Preliminary Draft of the General Budget of the EC for the Financial Year 1980* (hereafter *PDB 1980*), Brussels, Commission, June 1979, Vol. 7/A, pp. 99–102.

39 See Article 203, para. 9; see also pp. 87 ff. During the dispute on the 1979 budget the Parliament's adopted budget increased appropriations beyond the maximum rate of increase agreed by the Council. On 22 March 1979 the Council of Ministers made a 'declaration' that should this problem recur the Council would consider revising the maximum rate and try to identify priorities, or in the final analysis cut proportionately amended appropriations. The Dutch government dissented on the grounds that the formula threatened to undermine the budgetary powers of the European Parliament. For text see *Second Report ... of the House of Lords*, Session 1979/80, p. 7. In *PDB 1980* the Commission proposed a much greater increase in appropriations for non-compulsory expenditure (43 per cent for commitments and 26 per cent for payments) than the calculated maximum rate of 13·3 per cent, Vol. 7/A, pp. 4 and 98.

40 See Strasser, op. cit., pp. 32–3. Payment appropriations relate to expenditure actually to be incurred during the stated financial year, part of which may be in settlement of commitments previously made. Commitment appropriations state the limits of resources to be pledged in the current financial year, part of the payment of which may be spread over subsequent years. The distinction originated in the Euratom Treaty, Article 176, but was not applied to other areas of expenditure until Regulation 76/919/ECSC, EEC, Euratom of 21 December 1976 was approved. Even then it was agreed as applicable only to some areas of expenditure, with many other areas dependent on a single set of 'undifferentiated' appropriations for payment during the current year. In addition special exemptions from the payment rules apply, for example, to EAGGF (Guarantee) in case of difficulties in disbursing actual payments within the financial year. See *PDB 1980*, Vol. 7/A, p. 133.

41 See, for example, *Global Appraisal of the Budgetary Problems of the Community*, COM(79)85 final, Brussels, Commission, 12 March 1979.

42 On the EUA see above, note 26. See also Financial Regulation of 21 December 1977 *OJ* L356, 31 December 1977; *Bull. EC*, 10/1978, pp. 118–9; Strasser,

op. cit., pp. 38–48. Currently the EUA is based directly on the ECU, derived similarly from a basket of national currencies. By Regulation (EEC) No. 652/79 of 29 March 1979, *OJ* L84, 4 April 1979, agricultural expenditure is also expressed in ECU, though through a complex formula that does not alter *per se* price levels in national currencies.

43 For an explanation of MCAs see R. W. Irving and H. A. Fearn, *Green Money and the Common Agricultural Policy*, Centre for European Agricultural Studies, Wye College, 1975. For their relevance to the receipts of member states, see *Second Report ... of the House of Lords*, Session 1979/80, pp. xx–xxii, xxvii–xxviii and 10–11. It is important to note that in 1975 the UK was not a recipient from the EC budget largely as a consequence of the way MCAs were then operating.

3 The Decision-Making Process

The study of decision-making on financial issues offers particularly interesting insights into the political process of the EC. The budgetary process itself is rather tightly structured, in that it depends on precise rules about the operations of individual institutions and their interactions. Thus decision-making is much more systematically organised than in most other areas of Community activity. The annual budgetary cycle is the pivot around which the process revolves, and this depends on a definite timetable with built-in deadlines, which in turn circumscribe negotiations and generate strong pressures for agreement. Though there are mechanisms which permit provisional budgets to operate, if new decisions have not been produced by the due date, in general the timetable has been followed fairly rigorously. Deadlines also feed in from other specific provisions in the Treaties and the subsequent legislation of the EC, and thus add to the constraints on negotiations.

Member governments have been prepared to allow the establishment of strict procedures and to sanction transfers of political power that are more *communautaire* in character than almost any other area of Community responsibility. Thus the budgetary process stands alongside the bargaining on the CAP and the negotiation of external trade agreements as a prime example of a highly articulated Community process. In particular the evolution of decision-making on finance has generated significant innovations of a constitutional kind, notably in the extension of budgetary powers to the European Parliament. At the same time, however, analysis of this process also reveals major weaknesses in decision-making. Community institutions in practice do not facilitate discussion of the budget and other financial instruments as an aggregate whole. Instead decisions tend to emerge *seriatim* and then to be sewn together into a patchwork. The scope for contrary if not contradictory directions in financing is considerable, and the political choices that lie behind the raising and spending of money are rarely confronted openly.

The Provisions of the Treaties

Each of the basic Treaties sets out distinct rules and procedures to govern the budgetary process. Gradually, however, the Merger Treaty of 1965 and the Amending Treaties of 1970 and 1975 have imposed a unity of approach on the three Communities. The Treaty of Paris of 1951 largely conferred financial powers on the High Authority of the ECSC, though the administrative budget was determined by a special committee consisting of the presidents of the four ECSC institutions. Otherwise, however, the High Authority had only occasional recourse to the member governments in determining how money should be spent. By contrast the two Treaties of Rome charged the Commissions of Euratom and the EEC with drafting a preliminary budget, but it was the Council of Ministers that determined the final version. All three Treaties gave the Parliament the rights to be consulted and to propose amendments, but without any guarantee that notice must be taken. The Merger Treaty of 1965 established a single General Budget that incorporated most of the separate programmes of the three Communities and subjected them to the same rules and procedures, though the original Treaties continue to govern the operating expenditure of ECSC and part of Euratom's expenditure.[1]

Far more substantive changes were introduced in 1970 under the Amending Treaty that established 'own resources'.[2] Most important, this gave the European Parliament a positive right to suggest alterations to the draft budget, either as amendments in the case of *non-compulsory* expenditure, or as modifications in the case of *compulsory* expenditure. It also compelled the Council to take a formal decision on the points made by the Parliament and to enter into a dialogue with it on disputed points through what has become known as a *concertation* procedure. Furthermore, it was the President of the Assembly who was to declare the adoption of the budget. Thus under this Treaty the Council and the Parliament became the two arms of the EC 'budgetary authority'.

A second Amending Treaty on finance was agreed in July 1975 which further refined the procedures. Most important, it extended the role of the Parliament and created the Court of Auditors (the latter will be discussed in Chapter 4). The main effects of the 1975 Treaty in terms of decision-making were: first, to strengthen Parliament's powers of amendment and modification; and secondly, to give the Parliament the option of rejecting the budget as a whole. The consequence of these changes has been cumulatively to tilt the balance of decision-making on budgetary questions towards the European Parliament, in the sense that the final say of the Parliament now represents a significant lever of political influence. The reason for its acceptance, even by the several governments that have consistently expressed reservations about any extension of the Parliament's powers, is that once the decision had been taken to establish 'own resources' it proved impossible to resist the logic of allowing a degree of Parliamentary influence;[3] otherwise there would be no democratic check on Community taxation and expenditure. Equally it can be argued that in the absence of a firm agreement on 'own resources' the Parliament might well still lack any definite powers beyond the rather limited provisions of the basic Treaties.

The Budget Procedure

The budget procedure, as it currently operates, is now wholly based on the 1975 Treaty, supplemented by specific rules set out in the most recent Financial Regulation of December 1977.[4] This latter prescribes in detail the principles and rules that govern the General Budget of the three Communities, expands some of the technical points mentioned in the Treaty and clarifies aspects of the relationships among the various institutions. In addition, by mutual agreement within and between the relevant institutions, various refinements have been added for all practical purposes, though they do not have the full force of the Treaty. In particular the timetable for the annual preparation of the budget has been brought forward, the distinction between *compulsory* and *non-compulsory* expenditure has been clarified, and the *concertation* procedure between the Council and the Parliament has evolved after some four years of experience. Grey and disputed areas still remain, some of which have been the subject of protracted debate, notably in connection with major differences of opinion between the Council and the Parliament. A recent example was the dispute over the 1979 budget, when in December the Parliament declared 'adopted' a budget that contained higher allocations for non-compulsory expenditure, especially on the RDF, than the Council had collectively agreed. The most important of these concerns amendments that raise total expenditure in excess of the declared *maximum rate* of increase, to which we shall return. This apart, the main stages of the budgetary procedure are as follows:

(1)   There is an initial discussion in April by the Council of Ministers (both foreign and finance) on the strategy for next year, with a paper from the Commission.

(2)   The preliminary budget is drafted by 15 June by the Commission and the Budget Council in July establishes the draft budget.

(3)   The Parliament formulates amendments and modifications in October and these are considered by the Budget Council on 20 November.

(4)   After *concertation* between Parliament and the Council, the Parliament on 14/15 December decides whether to accept or reject the budget as a whole.

The Commission

The Commission's responsibility in the budgetary process is to propose, resting on its collective expertise and its function as the manager of EC finances. The largest proportion of the budget by far is that which is spent by the Commission, as Table 3.1 shows, and the Commission also operates the 'own resources' under the Council Decision of April 1970. Thus the Commission aggregates its own budget together with the administrative budgets of the other institutions into a single document presented to the Council. Within the Commission this responsibility devolves on the Commissioner responsible for the budget and Directorate General XIX, and they collate information from other Directorates General (DGs) to present a co-ordinated document for approval by the college of Commissioners.[5] It is, however, important to note that there is a degree of tension in the Commission between the principle of collegiate responsibility as exercised by the thirteen Commissioners collectively and the decentralisation of functions to individual Commissioners and DGs. This is compounded as far as expenditure is concerned by the different rules that govern individual spending programmes, by the lack of flexibility in agricultural expenditure on the prices policy and by the enthusiasm of each DG to increase its budget appropriations in order to develop policies. Inevitably, therefore, the collective Commission view on the budget represents an internal compromise that reflects the different constraints and calls on EC financial resources. Invariably the outcome is a proposal for substantial increases in expenditure and correspondingly larger calls on revenue. This is not to say that the Commission simply accommodates internal demands by massive increases in proposed expenditure in an irresponsible fashion. A marked characteristic of the Commission's approach is, however, to seek steadily to enlarge the scope and volume of EC financing, not least because of the Commission's awareness that the Council's view is likely to be more restrictive, an awareness that encourages a high initial bid from the Commission.

Over the years there has been a shift in the way the Commission approaches the budget. In the early years of the EC the tendency of the Commission was to press for a maximalist budget in order to maintain the pressure on governments to extend the budget as far as possible. During the period of economic prosperity of the 1960s this approach was not seriously at odds with the economic environment. Recently, however, the Commission has become increasingly sensitive to the political and economic constraints within which it has to work. Consequently its treatment of the budget has become more cautious than the bravado of the mid-1960s, which generated proposals of a radical kind for budgetary expansion and also a large say in financial allocations for both the Commission and the Parliament. The Commission remains a determined advocate of a more *communautaire* approach to the budget in terms of both general principles and specific provisions. However, this stance is increasingly tempered by the need to recognise the political views of member governments.[6] There are, in any case, limits to what the Commission can propose by way of extra expenditure in any annual budget-cycle, since in many cases appropriations can be used only on the basis of decisions by the Council that certain kinds of expenditure are permissible. The Commission recognises

Table 3.1 Distribution of the Budget Among the Community Institutions

Total appropriations for commitments[1]

Institutions	Appropriations 1978 Amounts	%	Preliminary Draft Budget 1979 Amounts	%	Increase for 1979 over 1978
Parliament[2]	100,424,612	0·79	111,089,905	0·75	+ 10·62
Council[2]	97,117,702	0·76	104,942,900	0·71	+ 8·06
Commission	12,478,076,795	98·23	14,617,144,575	98·30	+ 17·17
Court of Justice	17,332,920	0·14	21,267,200	0·14	+ 22·70
Court of Auditors	9,982,055	0·08	14,762,305	0·10	+ 47·89
TOTAL	12,702,934,084	100·00	14,869,206,885	100·00	+ 17·08

Notes:
1 Taking account of the Second Supplementary Budget for 1978.
2 The total appropriations for commitments are equal to the sum of the appropriations for commitment of the differentiated appropriations, i.e. of the appropriations for which a distinction is made between appropriations for commitment and appropriations for payment, plus the amount of the non-differentiated appropriations.

Total appropriations for payments[1]

Institutions	Appropriations 1978		Preliminary Draft Budget 1979		Increase for 1979 over 1978
	Amounts	%	Amounts	%	
Parliament	100,424,612	0·81	111,089,905	0·79	+ 10·62
Council	97,117,702	0·79	104,942,900	0·75	+ 8·06
Commission	12,137,797,303	98·18	13,807,451,530	98·21	+ 13·76
Court of Justice	17,332,920	0·14	21,267,200	0·15	+ 22·70
Court of Auditors	9,982,055	0·08	14,762,305	0·10	+ 47·89
TOTAL	12,362,654,592	100·00	14,059,513,840	100·00	+ 13·73

Note:
[1] The total appropriations for payments are equal to the sum of the appropriations for payment for differentiated appropriations plus the sum of the non-differentiated appropriations.

Source: Preliminary Draft General Budget of the European Community for the Financial Year 1979, p. 5, Tables 1 and 2.

both the budgetary and the political consequences of a budget so heavily skewed towards agricultural expenditure, yet the balance cannot be altered without either a change in the CAP or the adoption of programmes for other sectors. Thus the specific contribution of the Commission to the budgetary process cannot be separated from its parallel roles in making proposals on agriculture or on new Community instruments. Here the record of the Commission has been one of repeated emphasis on new instruments, as for instance in pressing for the establishment of the RDF.[7] More recently the Commission has begun to press proposals for loan facilities such as the Ortoli facility, since these escape the now-tight constraints on budgetary expansion.

In organisational terms, however, there are obstacles in the path of the emergence of a coherent budgetary strategy. The Commissioner responsible for the budget has an important portfolio, but with a status that is far less than that of a finance minister within a national European government, who normally combines (with the exception of Italy) budgetary responsibility with a broader financial role. In the current allocation of Commission portfolios a different Commissioner is responsible for Economic and Financial Affairs (through DG II), and for Credit and Investments (through DG XVIII), and yet another for Tax Harmonisation (through DG XV).[8] The more accurate parallel is to be drawn with the American system of government in which responsibilities are not dissimilarly divided. In both cases this reflects the fact that purse-strings are held elsewhere, in the EC case largely by the Council with some Parliamentary influence and in the US case by Congress. It also illustrates the division of portfolios to Commissioners so that individuals of different nationalities have roughly comparable status. This dispersal of authority combined with the inevitable tension between the manager of the budget and those responsible for actual spending programmes limits the impact that the Budget Commissioner and DG XIX can make. Suggestions have been made, promoted notably by the German government in 1974/5, that the Budget Commisssioner's portfolio should be translated into something more akin to that of a finance minister, but no such reform has yet been made.[9] A major change, however, in the Commission appointed in 1977 was the clear separation of the budget portfolio from any major spending portfolio.

The other outstanding feature of the Commission's attitude to the budget is that it remains the guardian of Community orthodoxy. In the context of the debate on equity it is important to recall that the Commission has until very recently held firmly to a purist view of the bases of 'own resources', especially in its insistence that levies and duties cannot and should not be regarded as separate national burdens. Equally the Commission has strongly resisted being drawn into the debate over the national balance sheets of contributions to and receipts from the EC budget. However, in response to the attitudes expressed by some member governments, and in recognition of the political and economic arguments that lie behind them, the attitude of the Commission has shifted. This was most evident in the document on future sources of revenue, 'The Way Ahead', in which the Commission explicitly recognised the problem of equity, and sought, albeit tentatively, to indicate some lines of escape from the dilemma that it poses. By contrast there has consistently been a strand in Commission thinking that has advocated a redistributive character for the budget, and this explains in part the Commission's dislike of national quotas for EC spending programmes.

The Council of Ministers

Within the Council of Ministers there is a special Budget Council composed of appropriate ministers from national finance ministries. Its sessions are prepared in detail by a Budget Committee of finance officials, and passed to the Council via the Committee of Permanent Representatives. These organs are reponsible for the detailed discussion each year of the preliminary draft budget submitted by the Commission, and for conducting the dialogue with the European Parliament over amendments and modifications. The structure of this Council process is tightly knit, and in most years enables the Council to conclude its budgetary deliberations within the timetable set by the 1975 Treaty. Part of the explanation for this relative 'efficiency' compared with other areas of Council negotiation lies in the system of deadlines, which has built into the process a regular momentum and rhythm. However, it is also highly significant that budgetary negotiations are alone within the Council in making systematic use of majority voting. The Treaties provided for majority votes specifically for the budget, and even included different weighting for individual expenditure areas, as Table 3.2 shows. Moreover, the practice has been accepted and followed by all member states, even those with reservations about the principle of voting versus unanimity. Indeed, the use of voting is further complicated by the rules of the 1975 Treaty, which require some proposals to attain a qualified majority in favour for their adoption (such as to alter the Commission's initial proposals or the Parliament's modifications to *compulsory* expenditure), while other proposals stand unless a qualified majority rejects them (as for the Council's response to Parliament's amendments to *non-compulsory* expenditure). The consequence is that member governments have to follow carefully thought-out tactics in determining how to vote on the individual items in the budget. Problems arise both on rather technical points and on the broader political issues, where the effect of a government casting its vote a particular way may be related not just to its view of the substantive proposal but also to the implications for the relationship between the Council and the Parliament. Further it must be noted that the *concertation* dialogue between Council and Parliament requires active involvement by the Council Presidency, with the corollary that the stance and style of the presidency is affected to some extent by which government is in the chair at the time.[10]

Though the Budget Council carries out the detailed negotiation on the annual budget, it does not operate in a vacuum, nor is it the Council of last resort. Where there are important conflicts of view among governments these may be referred up the Council hierarchy and ultimately to the European Council. Particularly in instances where the source of difficulty is a proposed increase in total expenditure, heads of government are likely to be brought in to arbitrate. However, this presents a major legal problem in that the decisions of the European Council do not have force until endorsed by the Council of Ministers properly constituted under the Treaty framework. The constraints of this are major, especially given the pressures of the very tight timetable for the completion of the budgetary procedure.[11]

The other central difficulty of the Budget Council is that it cannot always determine the levels of expenditure for particular programmes, but only aggregate the figures after 'spending' Councils have reached their individual decisions. Thus the broad levels of expenditure are determined by the appropriate technical

Table 3.2 *Distribution of Votes in the Council of Ministers*

	Admin. budgets	EDF	Social Fund	Euratom research	General budget
		Community of Six			Community of Nine
Belgium	2	11	8	9	5
Denmark	—	—	—	—	3
Germany	4	33	32	30	10
France	4	33	32	30	10
Ireland	—	—	—	—	3
Italy	4	11	20	23	10
Luxembourg	1	1	1	1	2
Netherlands	2	11	7	7	5
United Kingdom	—	—	—	—	10
Qualified Majority	12	67	67	67	41

Sources: Treaty establishing the European Economic Community; Article 148; Treaty establishing the European Atomic Energy Community, Article 118; Treaty establishing the European Coal and Steel Community, Article 28; and the Implementing Convention on the Association of Overseas Counties. Article 7.

Notes:

1 The voting figures for the General Budget of the EC of Nine supersede those in the previous three columns.

2 Separate procedures apply to the operating budget of the ECSC (Articles 53–6), requiring unanimity and others needing the 'consent' of the Council or a two-thirds majority. In these cases special power is given (Article 8) to states that produce at least one-eighth of the total output of coal and steel in the EC: Germany, France and the United Kingdom.

or specialist Councils that negotiate on policies. The difficulties are most acute in the case of agriculture, where the rate of budgetary expenditure is a function of the decisions on prices, made in the Council of Agriculture Ministers, and also a consequence of the volumes of agricultural production and trade. Thus the process does not permit the Budget Council to set anything equivalent to 'cash limits' for all areas of expenditure. In practice limits are imposed *de facto* on areas of spending outside agriculture, but often in a residual way rather than as part of a coherent strategy. Major factors here are the requirement that the budget must balance and the difficulty of introducing supplementary budgets to raise extra revenue. A further complication is that any decisions on new expenditure that precede their inclusion as budgetary appropriations are also made in appropriate specialist Councils, or occasionally in the General Council of Foreign Ministers. Members of the Budget Council are not formally involved in the negotiations at the Community level and their influence is confined to the pressures put by finance ministries on other departments in individual capitals or to the contributions of finance officials to discussions within 'spending' councils. Recognition of this has emerged in the various proposals for reform of the Council of Ministers. Most important, since 1976 the practice has been instituted of holding a Joint Council of Foreign and Finance Ministers in April.[12] Its function is to look at the broad shape of revenue and expenditure in strategic terms, before detailed discussion begins. Though the first meetings of this Joint Council made little identifiable impact, there is now some evidence that awareness of the interrelationships among the different components of the budget has improved, though not yet to the extent of producing a thoroughly coherent approach.

The European Parliament

From the outset the acquisition of budgetary powers has been an important symbolic issue for the European Parliament, as for the development of national parliaments historically.[13] During the mid-1960s the failure to win the approval of the French government for an extension of parliamentary influence over the budget helped to trap the Parliament into the vicious-circle argument as to whether it should press first for more powers or for direct election. With the adoption of the 1970 Amending Treaty and its reinforcement by the 1975 Treaty the circle was broken, thus facilitating agreement in 1977 on direct elections. A process has followed whereby the scope of Parliament's powers has expanded both *de jure* and *de facto*, to the point where members of the old selected Parliament have come to regard the passage of the budget as an arena for regular confrontation with the Council of Ministers. It is widely assumed that a Parliament based on direct elections will expect to flex its muscles by maximising its influence on the budget. Moreover the explicit recognition in the Treaties of the Parliament's role in the budget has enabled the Parliament to establish its position in other areas of Community activity too. Thus there is now more regular consultation, including a specific *conciliation* procedure between the Council and the Parliament on those legislative matters that carry financial implications, again a major extension of parliamentary influence.[14]

The budgetary process involves the Parliament both in rather intensive and detailed scrutiny of the draft budget and also in the broader debate of general issues. Most of the detailed scrutiny devolves on the Budget Committee, which is among the more prestigious and diligent Committees of the Parliament.[15] The Budget Committee produces reports that comment in considerable detail and then form the basis of any amendments or modifications submitted to the Council. The report is debated in a plenary session, from which a definitive document goes to the Council. From a political point of view the most interesting phase occurs when the Council returns its own final version of the budget to the Parliament for adoption. At this stage the members of the Parliament are called on to make a political judgement on whether to press their case on points of disagreement. Here the record of the Parliament so far has been unequivocally to seek to ensure that its budgetary powers are widely rather than narrowly exercised, in order to disprove the contention of those who have seen the Treaty provisions as conveying nominal rather than substantial influence. Thus the Parliament tested the 1970 Treaty at the first opportunity, when in March 1975 a Supplementary Budget was proposed to provide appropriations for the new RDF.[16] The bone of contention then was whether regional expenditure should be classified as *compulsory* or *non-compulsory*. Eventually the Parliament's point was recognised in that in practice convention has now established a rather narrow definition of *compulsory* expenditure, confined to the bulk of the EAGGF and finance committed under a firm international agreement, to which the EC are party.

The Parliament also chose to press the Council on its right to amend *non-compulsory* expenditure and on the limitation to the *margin of manoeuvre*. The Council failed to reject by the due date an amendment from the Parliament greatly increasing the size of the RDF. The Council expressly did not agree to an accompanying increase in the *maximum rate* and thus in the *margin of manoeuvre*. In December 1978 the Parliament adopted the budget for 1979, including increases in expenditure but ignoring the Council's express reserve against breaching the *maximum rate* provisions.[17] The consequence was to provoke a major dispute within the Community institutions. The Commission held that the budget had been lawfully adopted, even though it departed from the precise requirements of the Treaty with regard to the *maximum rate*. The Council was divided between those who accepted this, while criticising Parliament's handling of it, and the three governments of Denmark, France and the UK, which held that the Parliament had exceeded its Treaty powers. The UK took this position on a point of principle about the powers of the Parliament, even though it stood to benefit from a larger RDF. This latter began to make payments to the Commission, in accordance with Article 204 (EEC) on 'provisional twelfths' consisting of monthly payments of one-twelfth of the previous year's budget, on the grounds that a proper budget for 1979 had not been established. After several weeks of tense negotiations in which there was a possibility of reference to the Court of Justice, a compromise emerged around a Supplementary and Amending Budget eventually adopted by the Parliament on 25 April 1979. The effect of this was to increase appropriations for the RDF substantially by comparison with the majority view of the Council. It also introduced additional appropriations for the new interest rebate scheme for the 'less prosperous' members of the EMS, which had been agreed only at the European Council of December 1978. The Council discussion of this included a declaration by eight member states, with the Netherlands dissenting, on a formula to guide the *maximum rate* provision in subsequent years.[18]

Three major points emerge from this episode. First, it is clear that the Treaty provisions do not constitute a watertight guide to the budgetary procedure, particularly as regards its final stages. Even though the Council may have closed the door on the issue immediately at dispute in December 1978, several other horses are waiting to bolt from the stable. These include the precise definition of the Parliament's allowed *margin of manoeuvre* in the volume of *non-compulsory* expenditure, and the legitimacy of the insertion of new headings for appropriations to the Commission in advance of firm Council decisions. The 1978 dispute illustrated the variety of authoritative interpretations that can be posited for particular Treaty provisions. Secondly, the Parliament as yet has substantial influence only over *non-compulsory* expenditure and cannot therefore make significant inroads into agricultural expenditure. Its only options on the latter are to propose modifications to decrease appropriations or to reject the budget as a whole. However, its record to date has been consistently to support increased funding for other sectors including notably the RDF and food aid. There is thus a close correspondence between parliamentary and Commission thinking on the shape of the budget that differs markedly from the views of many member governments. Its powers of amendment give the Parliament in effect the opportunity steadily to alter the shape of the budget, albeit at a fairly gentle pace. Thirdly, the powers of Parliament combined with the use of majority voting in the Budget Council give scope for visible interaction between some parliamentarians and some member governments to alter the contents of the budget. Thus a government such as the Italian which supports both a stronger Parliament and greater regional expenditure can exploit parliamentary support in Council bargaining. This has introduced an important new factor into the process of negotiations among member governments, and begun to force out into the open some of the more overtly political dilemmas about the evolution of EC finances. How much further this extends will depend in large measure on how the Parliament chooses to approach the budget in the future.

Decisions on Other Financial Instruments

The discussion so far has centred on the interplay among the institutions in respect of the annual budget. However, the various other financial instruments of the EC are separately handled outside the tight framework of the budgetary procedures. Thus the European Investment Bank is an autonomous body that is not directly subject to influence from any other EC institution. Its governors and managers include individuals nominated by the member states and by the Commission but not as formally delegated representatives. Attempts have been made to establish a closer dialogue between the EIB and the Commission in particular, in order to achieve greater coherence in decisions about the use of the various financial instruments.[19] However, the concern of the EIB to protect its independence has kept it at a distance from the rest of the Community process. The workings of the new Ortoli facility will be an interesting test of the progress made in increasing liaison between the EIB, the Commission and the Council. Similarly, decisions about borrowing and lending are made separately from the budget, though the Parliament has expressed its concern to be able to discuss these activities in parallel to the budget proper.[20] Given the recent trend in the EC to emphasise the role of loan and credit facilities rather than orthodox budgetary expenditure, the untidiness of the current arrangements for reaching decisions on them may prove an obstacle to their development into financial resources that are significant as instruments of coherent policies. This is a particular issue in the context of the current debate about resource transfers and convergence. Quite apart from the economic debate as to whether loans actually can constitute a significant transfer of resources, there is an important political question of whether the EC process is capable of integrating its loan and credit facilities into a broader strategy of economic and financial policies that would reinforce the instruments already available in the budget.

The Characteristics of Bargaining

Budgetary questions have been amongst the most hard-fought in Community negotiations. They raise substantive issues in their own right, but they are also a touchstone of broader attitudes towards the development of the EC. Arguments about money have lain at the heart of many of the most contentious disputes in the history of the EC, notably the 1965/6 crisis, the renegotiation by the British government in 1974/5 of the terms of UK membership and the recent conflict between the Parliament and the Council over the 1979 budget. Debates about contributions to and receipts from EC finances inevitably highlight national attitudes towards, and expectations of, the EC. The mere existence of Community funding on a now quite significant scale means that there are substantial groups within the EC who have vested interests in the preservation of current expenditure programmes. Equally it provokes claims from other groups for access to a share of EC resources. These difficult issues are rendered still more complex by the linking of financial questions to the broader issues of institutional powers and the balance between Community and national authority.

Yet in spite of the great sensitivity that attaches to these issues, the budgetary arena has generated major innovations in

Community decision-making. The original decision of the six founder members to establish a commonly financed policy for agricultural prices was, after all, a remarkable achievement of Community solidarity. The agreement in 1969/70 to create the 'own resources' represented a substantial strengthening of the Community idea. The gradual addition of new expenditure programmes and financial instruments, modest though most of them are, indicates the extent to which the EC have become accepted as a vehicle for collective action. The progressive increase in the role of the Parliament has added a major new dimension to Community negotiations that in turn changes the context in which the Commission and the Council operate.

This said, the crucial question for the political analyst is whether the accretion of Community responsibility for collective financing has a cumulative momentum strong enough to establish a consensus for its further development. Many of the signs currently in evidence appear to suggest that member governments at least have growing reservations about the logic of a continued increase in Community financing on the model so far established. Moreover, the issues that now lie at the heart of the budgetary debate have become explicitly more political and more contentious than at any period since the 1965/6 crisis. Once the questions of equity and redistribution have been brought into the open, as they have in recent debate, it becomes less easy to patch and to accommodate without tackling these questions head on. Here the record to date offers less than convincing evidence of the capacity of the EC as a locus for resolving issues of such magnitude. Only in the initial compact among the six founder members did governments collectively embrace commitments and reach a consensus on a comparable scale. Yet the debate over EC finances impinges on so many other areas of Community activity that it is difficult to envisage substantial progress being made on common policies without a concomitant accord on their budgetary implications.

Notes: Chapter 3

1   Treaty establishing a Single Council ..., Articles 20 and 21.
2   Treaty amending certain budgetary provisions of the Treaties establishing the European Communities and of the Treaty establishing a Single Council and Single

Commission of the European Communities, agreed on 21 April 1970 in Luxembourg, *OJ* L2, 2 January 1971.

3   See D. Coombes and I. Wiebecke, *The Power of the Purse in the European Communities*, London, RIIA/PEP, 1972.

4   Financial Regulation of 21 December 1977 applicable to the General Budget of the European Communities, *OJ* L356, 31 December 1977.

5   Daniel Strasser, *The Finances of Europe*, New York, Praeger, 1977, pp. 50–7; and Christoph Sasse, Edouard Poullet, David Coombes and Gérard Duprez, *Decision Making in the European Community*, New York, Praeger, 1977, pp. 166–9.

6   See, for example, Christopher Tugendhat, 'Problems of Community budgeting', *The World Today*, August 1977, pp. 287–94.

7   See, *inter alia*, *Report on the Regional Problems of the Enlarged Community* (the Thomson Report), COM(73)550, Brussels, Commission, 5 May 1973.

8   See H. Wallace, W. Wallace and C. Webb (eds), *Policy-Making in the European Communities*, Chichester, Wiley, 1977, pp. 52–8.

9   The German proposal was endorsed by the British government in *Financial Control in the European Community*, Cmnd 6360, London, HMSO, December 1975, para. 6.

10 See G. Edwards and H. Wallace, *The Council of Ministers of the European Communities and its President-in-Office*, London, Federal Trust, 1977, esp. pp. 44–9.

11 The meetings of the European Council do not necessarily take place on dates that fit with the requirements of the budget procedure. Part of the explanation for the eruption of the 1979 dispute was that the European Council reached a view on the Parliament's amendments to the RDF after the date when the Council should have communicated its views to the Parliament.

12 See *Bull. EC*, 4/1976, p. 70, on the first such meeting.

13 See Coombes (ed.), *The Power of the Purse*, London, Allen & Unwin, 1975.

14 Conciliation derives from a Joint Declaration by the Commission, Council and Parliament, *OJ* 1975 C 89/1. For a discussion see John Forman, 'The conciliation procedure', *Common Market Law Review*, 16, February 1979, pp. 77–108, which argues that recently the procedure has become more influential. Important examples include the decisions on the creation of the RDF, the 1977 Financial Regulation, the setting up of the Ortoli facility and the EC food aid programme.

15 The Budget Committee in practice takes the leading role in both the budgetary and conciliation procedures, though other subject committees comment on the relevance of the budget to their specialist areas, and play a more important part in commenting on the substance of financial legislation subject to conciliation.

16 See Wallace, Wallace and Webb, op. cit., pp. 158–9; and Forman, op. cit., pp. 92–3.

17 *Second Report ... of the House of Lords*, Session 1979–80, pp. ix–xii.

18 See above, pp. 67–8; *Second Report of the House of Lords*, Session 1979–80, p. 7; and *Agence Europe*, 22 and 23 March 1979.

19 In 1975 the Council of Finance Ministers and the Commission began to increase their efforts to involve the EIB more directly in discussions of policy, though the EIB 'reacted in a very negative fashion' (*Agence Europe*, 3 July 1975). In practice, however, the EIB has been far more susceptible to political influence indirectly on the choice of projects for support than it admits.

20 See Forman, op. cit., pp. 94–5.

4 *Management and Control*

An assessment of the politics of Community financing cannot rest simply on a study of the way in which budgetary decisions are made. Just as important is what happens to the money that has been appropriated for Community use. The economic impact of the EC depends in part on the way Community spending programmes are implemented. Though still relatively modest the budget has to be deployed satisfactorily if the case for a larger exchequer is to be won. The claims of the Commission to a greater share of political responsibility depend in part on the confidence of governments in its capacity to manage policies and the funds that support them. Thus the processes through which EC finances are managed and controlled contribute significantly to the judgements of the economists and politicians on their effectiveness. There are two dimensions to the framework for management and control: the first consists of the internal procedures for handling Community money, and the second comprises the external procedures for their scrutiny. In a system as complex as that of the EC it is inevitable that these procedures should be intricate. The division of competences between the Community and national levels both politically and administratively means that the responsibility for handling finances is dispersed among different agencies, and the lack of a strong central Community executive means that the EC component is itself fragmented. Thus the management of spending programmes, with the exception of a tiny minority of direct EC projects, depends on collaboration between Community and national bodies. This enlarges the scope for difficulties in implementation, simply because of the variety of constraints and obstacles in the long chain of procedures. Equally thorough scrutiny of the deployment of resources is problematic, when it is so difficult to pin down precise responsibilities. In the early years of the EC far more attention was given to the need to define managerial responsibilities than to the establishment of rigorous control procedures. More recently efforts have been made to introduce more systematic scrutiny, partly in response to evidence of maladministration and fraud. They also reflect the altered climate in which EC finances now operate, a climate in which value for money has become more highly prized, constraints on the rate of budgetary expansion have become more pressing, and public concern over the utilisation of financial resources has been more loudly voiced.

Efficient Management

The Commission is the Community organ with the prime responsibility for implementing financial programmes, apart from the loan and credit facilities managed by the EIB or through the network of national central banks. However, the degree of independence accorded to the Commission varies from the few programmes which it manages entirely on its own to the funds which in practice are largely deployed by agencies within the member states. Thus the Commission autonomously administers its own information programme, and funds a small number of projects and studies, but these account for a tiny proportion of the budget. These are determined according to internal criteria without the direct participation of member governments. For the rest a variety of formulae has been adopted to govern the implementation of individual spending programmes.

In the case of agricultural expenditure on price guarantees fairly precise rules are set in the legislation and the financial interventions are carried out by the intervention agencies in the member states. To determine the detailed application of the rules the Commission and national administrations co-operate through a network of committees. Most important among these are the Management Committees for individual agricultural products, the chairmanship and secretariat of which are provided by the Commission and the members of which are appropriate experts from national ministries of agriculture.[1] They meet with great regularity and express opinions frequently by majority votes on proposals for Commission legislation which the Commission can then adopt unless a qualified majority opposes them. The principles on which this intensive co-operation rests are first that the Commission is the manager and the national administrators are its agents, and second that the managerial role is basically a technical one of adjusting the application of predetermined rules to meet the actual operations of the agricultural market. The first principle holds in that no discretion is left to national authorities, but the second is less clean-cut. In practice the detailed interpretation of the rules can and does give rise to policy judgements that have political implications. Thus questions such as the determination of export subsidies are handled in Management Committees, even though the political repercussions of selling cheap butter to the Soviet Union may be great.[2] This intrusion of overtly political factors into a primarily managerial process may in practice mean that decisions are derived from more senior authorities, but this still depends on the Commission rather than member governments, unless the latter can muster a qualified majority against the Commission's proposal.

In most other areas of EC expenditure the managerial responsibility is more explicitly divided between Community and national authorities, in different proportions for individual programmes. So, for example, the Social Fund is allocated to projects that meet the criteria established in the Regulations, but the criteria are not defined with complete precision and their application is discretionary rather than mandatory. Consequently, the administration of the Fund requires careful interpretation and the establishment of priorities. To this end the Treaty of Rome (Article 124) created a Fund Management Committee, 'composed of representatives of Governments, trade unions and employers' organisations' to assist the Commission. This provision reflected the interest of the drafters of the Treaty in engaging in the financial process not just governments but also the ultimate recipients of Community grants. This, combined with the opportunity to differentiate among the claims of individual member states, has permitted the Commission considerable scope in interpreting the general rules for the Fund's use, and has encouraged the Commission to engage in an active dialogue with economic and social interests about spending needs and priorities.

The approach adopted for the management of the Regional Development Fund is quite different. The RDF was created in March 1975 after protracted controversy over its size and its priorities, in a climate of considerable scepticism about the economic impact it would have, and of reservations about the granting of managerial discretion to the Commission. Consequently, the distribution of the Fund was tied to national quotas which automatically reduced the scope for an active Commission role in its administration. Also a Fund Management Committee was created, but composed solely of the Commission, in the chair, and representatives of the member states. Consultations with regional interests were included in the Regulations only as a vague possibility. The combination of national quotas and an intergovernmental procedure have meant that the Commission is left to perform a largely administrative or technical role, with little scope for building policy judgements into decisions on expenditure. The recent decision to free a small proportion of the fund (initially 5 per cent) from the constraint of quotas may permit the Commission greater influence, but only over limited resources.[3] The consequence of those arrangements is in effect to place the burden of managerial responsibility on member governments, since they in practice make the choices about which projects should go forward for RDF support, and since they contribute a share of the finance directly from national exchequers. But the independence of governments is circumscribed by the need to accommodate national projects within the criteria laid down in the Community legislation. The overall effect has been to disperse responsibilities for implementation in such a way that the gains of decentralisation do not appear to confer increased flexibility in the allocation of the Fund's resources, particularly given that within the member states decisions about their use are in turn subject to delicate negotiations among different ministries and agencies.

A second major constraint on the implementation of Community finances derives from organisational factors. The Commission has only slender resources of staff particularly at the policy-making levels.[4] In order for the Commission to perform its implementing duties thoroughly and to monitor systematically the operation of EC finances within the member states, it requires substantial expertise and experience of the different practices and procedures followed in nine different countries. Personnel are not available on the scale that would be necessary for these functions to be exercised rigorously. In practice, therefore, the Commission staff have to be selective rather than comprehensive in the attention they devote to their managerial responsibilities. These difficulties are compounded by the other organisational weaknesses of the Commission, weaknesses well documented in the literature.[5] Thus the organisational effectiveness of the Commission is defined by Michelman as determined by variables related to the degree of Commission authority, internal values, the nature of its constituency and the interactions with the policy environment. Structural and psychological factors also impinge on the way individual DGs pursue their tasks. Clearly organisational effectiveness can be tested empirically by the analysis of financial management, especially given that the outputs in this case are more measurable than in some other areas of Commission activity. None of the academic research on the Commission has examined in detail the implementation of EC finances. However, it is important to note that such studies as have been completed, notably the recent study by Hans Michelman, indicate that there are great variations in organisational effectiveness among the individual DGs. On Michelman's variables, for example, DG VI (responsible for agriculture) scores highly by comparison with DG V (responsible for the ESF). This reflects in part the greater scope for active policy-making allowed to DG VI under Community legislation, but equally it equips DG VI to pursue its financial responsibilities more effectively than is possible for DG V. The difficulty that flows from this is the vicious circle through which some DGs are adjudged not sufficiently effective to be allowed a greater managerial role, while the limitations on their autonomy make it harder for them to become more effective.

A third constraint stems from the fragmentation present within EC institutions. Broad financial decisions are taken separately by different compositions of the Council of Ministers and then implemented by separate DGs in the Commission or by other EC organs such as the EIB. Criteria for financial programmes are determined according to internal criteria, even though individual programmes have implications for each other. To take the clearest example, expenditure on regional policy rests not simply on the RDF, but on the impact of agricultural expenditure (both guidance and guarantee) on particular regions, on the activities of the ESF, especially under Article 5, and on the instruments of the EIB. For Community finances as a whole to be managed coherently and consistently in particular regions requires systematic co-ordination among different funds and programmes.

Yet the weight of the administrative system is to reinforce unilateral action by individual units, rather than to promote a co-ordinated approach. The theoretical collegiality of the Commission does not in practice apply to detailed implementing decisions, and the autonomy embodied in the statute of the EIB impedes a clear alignment of both policy objectives and specific decisions between the EIB and the Commission. This encourages inefficiency in the allocation of resources and diminishes the confidence of member governments in any substantial expansion of EC finances.

The various EC institutions have consistently paid lip service to the goal of improving the coherence of Community expenditure, but it is only recently that steps have been taken to establish more thorough co-ordination and appraisal of the collective opportunities for maximising slender resources. However, though this fragmentation riddles all the EC institutions, it is only within the Commission that active efforts are now under way to seek remedies. Basically these efforts consist of more systematic procedures for regular consultations among Commission DGs and with the EIB, and of the work of the new 'Task Force for the Co-ordination of Financial Instruments.'[6] Here it must be noted that these activities fall within the domains of the Commissioner responsible for economic and financial policy and of the Commissioner specifically concerned with the co-ordination of financial instruments, who also happens to be responsible for regional policy. They do not report directly to the Commissioner responsible for the budget. So far the most obvious achievement of these efforts has been to increase the reservoir of information on the interaction among the different financial instruments and to improve mutual awareness of the problem areas. In addition there have been steady efforts to concentrate the use of different EC financial resources in particular regions and to sharpen up the criteria for their use by considering more carefully their mutual impact. This still, however, falls short of a general strategy from which more positive action might flow. Within the Commission thinking has become a good deal clearer, though publicly policy has not yet altered radically.

Control Procedures

EC finances are subject to three different types of control and scrutiny, designed to ascertain that revenue and expenditure have been appropriately managed. Administrative controls operate through the largely internal procedures for monitoring income and expenditure, and rest primarily on the supervisory role of the Commission. Financial controls derive from a mixture of sources; internal controls within each EC institution, controls by national audit authorities, and most recently independent control through the EC Court of Auditors. Political controls depend on the efforts made by representative institutions to hold the executive accountable for their management of finances, an activity that is most appropriately the responsibility of the European Parliament, but also draws in potentially the Council of Ministers and national parliaments. Broadly speaking, all three types of control are directed at ensuring that revenue and expenditure have been incurred correctly within the terms of EC legislation, and that finances have been handled carefully, and appropriately. The investigation of these questions requires the application of suitable managerial procedures, the assessments of specialist auditors and the intervention of public representatives, in order to guarantee both independent checks and political responsibility. However, control procedures may simply have the narrow function of checking that the books balance and that items have been legitimately and correctly entered, or they may have one rather broader purpose of ascertaining that the management of resources has been efficient and effective. Only recently have control procedures within the EC context begun to extend from rather narrow and limited checks to formalised attempts to scrutinise finances more systematically and to evaluate their impact. This shift reflects an increased concern to ensure that EC resources are not misused, a concern that has grown as the size of the budget has expanded. The enlarged focus of control procedures has also led to a more explicit linkage between assessments of financial management and judgements about policy instruments and their objectives.

Administrative controls in one sense depend simply on the organisational capacity of the appropriate EC institutions to monitor and supervise the collection of revenues and the programmes or projects to which expenditure has been allocated. In principle, therefore, they require careful and responsible management, exercised in a systematic fashion. However, in practice the Community process has not facilitated the adoption of thorough administrative controls over finances. Part of the explanation for this lies in the limited personnel resources available to EC institutions, and notably to the Commission. Detailed control of expenditure programmes and of revenue collection has not carried a high priority in terms of the total workload, and staff have been concentrated instead on other activities. The emphasis has been on policy formulation rather than policy implementation, and deliberate attempts to curtail the bureaucratic expansion of the Commission have coincided with the period in which increases in expenditure and the creation of 'own resources' occurred. Obviously the several organisational weaknesses of the Commission have militated against the introduction of rigorous control procedures. A more fundamental problem, however, is the fragmentation of administrative and financial responsibility between the Community and national levels, particularly given the great variations in administrative and financial practices among the different member states. Thus the access of Commission staff to appropriate information and their capacity to evaluate it is heavily circumscribed. Nor can the operation of Community financing rest on assumptions of homogeneity or even compatibility of national management within the member states. Inevitably, therefore, the coverage of administrative controls is patchy. This permits administrative inconsistencies which affect the impact of Community finances on the different parts of the EC, which in turn influences the degree to which financial instruments achieve the purposes for which they were established. Clearly more systematic procedures at the EC level and more closely aligned national practices would reduce the scope for differential implementation of financial decisions, but these are not likely to emerge unless the problem is accorded a sufficiently high priority by both national and Community authorities.

Financial control is a cognate activity through which financial expertise is brought to bear on the scrutiny of revenue and expenditure. The exercise of financial control depends on both internal procedures within the organs responsible for managing financial operations and on external checks by appropriate auditing agencies. The object of these activities is to ensure that administrative, financial and legislative practices are sufficiently thorough and precise to encourage responsible management and to prevent abuse and deviation. Basically the pattern of financial controls in the EC context has historically been very untidy and has comprised a mixture of divergent national practices and rather tortuous procedures at the Community level. In some member states national audit authorities have automatically taken on an active responsibility for supervising the flows of EC finances through national agencies, whereas in other member states Community operations have either been perceived as extraneous or actually fallen outside their jurisdiction. Within the Commission DG XX is specifically responsible for financial control, and thus for authorising payments and validating accounts of income and expenditure. The exercise of financial control has rested primarily on the utilisation of rather laborious and time-consuming procedures that often magnify the importance of detail and neglect the objects of policy that lie behind the financial instruments.

The importance of these factors should not be underestimated, since they contribute significantly to the efficiency and effectiveness of EC financing. Misdirected efforts at financial control can cause delays in the utilisation of resources and build in inflexibility, yet at the same time lack the rigour to pick up instances of abuse either in the form of fraud or of evasion. Equally an inappropriate or insensitive pattern of controls can introduce frustration and misunderstanding between Community authorities and the agencies within the member states that collect or spend Community money. This in turn affects judgements about the coherence of financial management in the EC and about the capacity of the Community system to handle a larger range of income and expenditure.

Dissatisfaction with the existing patchwork of financial controls increased strongly during the early 1970s, particularly as the 'own resources' system was being introduced. The exposure of several cases of spectacular fraud and 'sharp practice' in the agricultural sector aggravated public concern at the apparent lack of attention to how EC money was actually spent.[7] Mechanisms had existed under the Treaties for independent investigation into EC finances through the ECSC Auditor and the Audit Board of the EEC and Euratom, but the remit given to these was too limited and their practices were too narrow to generate confidence in their observations. Consequently the Amending Treaty of 1975 included as a major innovation the establishment of a new organ, a Court of Auditors modelled approximately on the French *Cour des Comptes.* The intention was to create an entirely independent body at a Community level that would scrutinise thoroughly the various financial operations of the EC. Its tasks were threefold: to check first, the lawfulness and secondly, the regularity of the operation of both revenue and expenditure, and thirdly to ascertain that it had been subject to 'sound financial management'. The Court of Auditors was eventually set up in October 1977 in Luxembourg.[8] It is required to present annual reports on the previous year's accounts and may prepare special reports on its own initiative or at the request of other institutions. In addition it can offer observations on other financial proposals under consideration in the Community process. Its reports are published and are to be taken into account by the European Parliament when it discharges previous years' accounts.[9] Thus financial control from an independent body has received an elevated status and priority. The creation of the Court is an interesting example of institutional innovation, with the extra feature that nominations by the Council of members of Court have to be endorsed by the European Parliament.

The most important questions about the role of the Court are twofold: how thorough the audit process will be in investigating not just the lawfulness and regularity of EC finances but also the soundness of financial management; and secondly what the impact of the Court's observations will be on the way other institutions, both Community and national, handle EC finances. The Court's existence is too new for any clear assessment to be attempted yet, but there are preliminary indications of its own approach and of the response of other institutions to it. The members of the Court have been at pains to stress their distinction from the old Audit Board and to emphasise that they intend to focus on sound financial management as a serious objective. Thus they hope to make some impact on the policy framework, rather than simply to provide guarantees against fraud, evasion, and so on. Also the Court intends to be thorough in its coverage and it is already clear that this means identifying major weaknesses in the collection of 'own resources', an area previously neglected by the Commission.[10] The Court furthermore has the constitutional right to make 'on the spot checks' in the member states, the exercise of which is already beginning to compel more careful attention by national authorities to the way in which they handle EC finances. It would, however, be misleading to suggest that the Court will have a rapid and radical impact, since it has, first, to establish a coherent and integrated philosphy of its own role and, secondly, to contend with the fragmentation of Community responsibilities and the diversity of national practices. Deciding where the buck stops and ensuring that appropriate reforms of financial management take place are not going to be achieved overnight. Such aims depend too on how the observations of the Court are made to stick, and here achievement will depend on practice rather than simply on the prescribed legislative framework.

Three particular problems remain to be resolved. First, the early relationship between the Court and the Commission has shown some signs of strain as the Court has endeavoured to establish its own rights and as the Commission has sought to resist undue interference.[11] Secondly, the extent of access by the Court to national authorities that handle EC money has yet to be clarified. Thirdly, the nature of the Court's relationship with the European Parliament has yet to be determined. The Court has no direct constitutional responsibility to the Parliament, but the assumption of most commentators is that the Court's impact will be limited unless the Parliament chooses to throw its political weight behind the judgements of the auditors. Underlying any assessment of Court is the question of whether its efforts to improve financial management will in practice lead to a policy-making role by its indirect impact on those who operate EC finances.

Political Control

Fundamental to any budgetary process in a democratic system is the accountability of the executive to a politically representative body. In national terms this generally means the scrutiny by parliaments of revenue and expenditure, with the emphasis normally being placed on the expenditure side of the budget. In the EC context the application of this principle has been both difficult to define and of relatively low salience. The difficulty of definition stems partly from the lack of clarity over what constitutes the executive and partly from the great limitations on parliamentary scrutiny. While the Commission may in theory claim the status of an executive, its operations have been circumscribed by the policy framework generated by the Council of Ministers and impeded by its dependence on the agency of national administrations. In a sense the Council of Ministers is simultaneously executive in character, by virtue of its decision-making functions, and representative, as the body to express the views of member governments. In practice the Council of Ministers has devoted only limited attention to the control of revenue and expenditure, choosing rather to concentrate on the preliminary phase of the budgetary process. National parliaments in the member states can make only a limited contribution to the scrutiny of EC finances since they have no direct access to Community institutions. There is, however, arguably some scope for intervention by national parliaments to scrutinise the way in which their own governments handle EC finances.[12]

The brunt of responsibility for political control of EC finances lies firmly on the European Parliament. Historically this has not, however, been an area of prime importance. The Parliament until recently had only a limited constitutional role in this respect and in any case showed more regular interest in the initial passage of the budget. Occasionally parliamentarians pursued particular cases of alleged maladministration of EC funds and uncovered examples of fraud, but it was not until 1976 that the Parliament appended to the Budget Committee a Subcommittee on Control. In spite of the efforts of individual members the subcommittee has not made a significant collective impact on the management of EC finances. However, two developments have begun to put pressure on the Parliament to emphasise its involvement in financial control. First, the shift to direct elections and the accompanying interest in identifying further opportunities for an enhanced influence for the Parliament have encouraged members to exploit their available instruments for challenging the Commission and the Council. Secondly the Treaty provisions have been progressively amended to transfer the responsibility for the discharge of the accounts from the Council to the Parliament. Two conditions, however, must be fulfilled if this responsibility is to be exercised substantially and not merely symbolically. First, the role of the Subcommittee on Control would need to be strengthened within the parliament itself. A first step was taken in July 1979 when the new Parliament upgraded it into a full Committee on Budgetary Control.[13] Secondly, the Parliament collectively would have to develop a close relationship with the Court of Auditors in the sense of giving political support to those of the Court's observations that coincide with the views of parliamentarians. Otherwise the findings of expert financial controllers will not be translated into political pressures for improved management of EC finances or for increased attention to their effectiveness.

Notes: Chapter 4

1 There is little academic literature on management committees. The general reports on the activities of the European Communities list these committees and record the numbers of votes taken and their outcome. See, for example, *Twelfth General Report*, Brussels, Commission 1979, pp. 178–9.

2 As, for example, in March 1973, when the issue was raised and generated protracted political controversy after a 'technical' decision on export subsidies reached in the Management Committee on Dairy Products.

3 A consolidated text of the various regulations and guidelines was published in *OJ* C36, 9 February 1979. The quotas agreed for 1978–80 were Belgium, 1·39 per cent; Denmark, 1·20 per cent; Germany, 6·0 per cent; France, 16·86 per cent; Ireland, 6·46 per cent; Italy, 39·39 per cent; Luxembourg, 0·09 per cent; Netherlands, 1·58 per cent; UK, 27·03 per cent.

4 In reply to a question in the European Parliament, (no. 848/78) the Commission stated on 19 April 1979 that its establishment of staff was 10,722: 2,825 A grade (administrative); 2,570 B grade (executive); 3,765 C grade (clerical); 467 D grade (support staff); and 1,095 L grade (linguists).

5 See David Coombes, *Politics and Bureaucracy in the European Community*, London, Allen & Unwin, 1970; Hans Michelman, *Organisational Effectiveness in a Multinational Bureaucracy*, Farnborough, Saxon House, 1978; and Christoph Sasse *et al.*, *Decision-Making in the European Community*, chs 3, 4 and 5.

6 The Task Force was created in 1977 and staffed by officials from the DGs with financial responsibilities. Its main role so far has been to stimulate internal discussion rather than to publish its activities more widely.

7 In 1973, for example, the Budget Committee of the European Parliament discovered that between 100 and 130 million UA from EAGGF expenditure in 1970 was not accounted for. On 30 November 1973 the Commission announced the formation of a 'flying squad' to investigate frauds. See also *The case for a European Audit Office*, Luxembourg, European Parliament, 1973.

8 See *Second Report ... of the House of Lords*, Session 1979/80, pp. xxxii–xxxvi.

9 *Annual Report Concerning the Financial Year 1977 Accompanied by the Replies of the Institutions*, *OJ* C313, 30 December 1978 (the First Report of the Court of Auditors).

10 Annual Report ... 1977 ..., pp. 117–123.

11 There was an exchange of letters between the Commission and the Court, following publication of the *Annual Report ... 1977 ...*, over whether the Court had the right to comment on the replies of the Commission to the Court's preliminary observations.

12 See, for example, *Thirteenth Report of the Select Committee on the European Communities of the House of Lords*, Session 1976/7.

13 See *Agence Europe*, 21 July 1979. The first task facing the new committee was the consideration of a critical report from the old Parliament on the merits of granting a discharge for the 1977 accounts.

5 Current Developments and their Implications

The crucial question to ask about the future development of the budgetary process is whether the establishment of a Community budget on the analogy of national budgets lies at the centre of the process of economic and political integration. Most economists would accept that a single aggregated economy presupposes the existence of collective financial instruments, through which to exercise a degree of influence over its economic policies. Indeed the reasons for which some economists have criticised the current budgetary activities of the EC are either that already they diminish the scope for individual member states to pursue effectively individual national policies, or that they produce results that are positively detrimental to particular national or regional economies.[1] There is rather more disagreement among economists on which financial instruments should comprise the core of a Community budget. The chief purpose of the MacDougall Report was to define the essential components of public finance at the EC level during the various phases of the integration process. The different views expressed on this major issue reflect a lack of consensus about how far the establishment of a common market leads irrevocably towards the adoption of more closely knit economic and fiscal policies.

Briefly the argument of the MacDougall Report was that within a quasi-federal system a limited number of financial levers had to be pulled at the centre, and that this would then permit a significant impact to be made on economic performance throughout its territory, including the reduction of regional inequalities. Its reasoning was based on an examination of the financial arrangements in a number of existing federal systems. The conclusions that it drew from this were, first, that an interim 'pre-federal' budget might be effective with resources equivalent to $2-2^1/_2$ per cent of Community GDP, rising perhaps to 5–7 per cent (excluding defence expenditure), and secondly, that the concentration of expenditure should be on structural, cyclical, employment and regional policies. Such a development would exclude EC expenditure on the bulk of social and welfare policies and probably also stabilisation policies, such as deficit financing, but might include the more or less total transfer to the EC level of expenditure such as development aid, where there were potential economies of scale.[2] Such an approach runs counter to the current practice of the Community budget and indeed to the declared preferences of most member states, except those few, the 'less prosperous', who might derive positive gains from equalisation measures. It would require a major recasting of current expenditure priorities in the EC through the development of some of its presently small programmes and the addition of new programmes, with a commensurate reduction of some areas of expenditure, notably on agriculture.

A different dimension of this debate is highlighted in the old argument between 'economists' and 'monetarists' in the context of moves towards economic and monetary union. In the early 1970s this centred on whether steps towards EMU had to be taken in parallel on economic and monetary co-operation, or whether one could precede the other.[3] During the recent discussions on EMS and the associated 'concurrent studies' exercise this divide has taken on a new focus; it has moved away from simply expounding the virtues of aligning national economic policies to questioning whether actual transfers of resources are required to sustain a concerted monetary policy among member states with divergent economic trends. Here there is a sharp distinction between those who see resource transfers as a necessary corollary of monetary co-operation and those who regard them as a national rather than a Community responsibility.[4] The first view assumes a greater role for both EC budgetary and loan mechanisms, while the latter denies that limited steps such as EMS need have budgetary consequences. The findings of the MacDougall Report suggested that any such expansion of the EC budget might still look fairly modest in terms of total public expenditure or of Community GDP, though much of the statistical evidence on which their calculations were based pre-dates the current concern with indicators of economic divergence and convergence. There is a certain irony in the EC's beginning to confront issues of this nature at a time of changing attitudes to public expenditure within the member states, in so far as significant sections of political opinion are now advocating extensive reductions in the role of public finance.

To the political scientist the crux of the debate about the significance of the EC budget looks rather different. One important point of interest is that the establishment of a politically integrated system, with the attributes of governmental functions familiar at the national level, requires a sophisticated process for making common decisions and selecting common policies. This presumably would include a budgetary process as a central element. Thus attitudes towards the budget may be indicators of broader attitudes on the desirability of the EC emerging as an acceptable system of government. Reservations about the budget may conceal reluctance to accept a politically more cohesive Community, not least because an enlarged budget would impose heavier burdens and more stringent conditions about its deployment.[5] According to this view it is no accident that some of the most acute disputes among governments in the history of the EC have been generated by conflicts over the characteristics of the budget, nor that budgetary issues have always been entangled with arguments about their institutional implications.

In a narrower context budgetary politics in the EC about the shape of the budget and its utilisation have been very revealing of both the objectives of member governments and the capacities of EC institutions. The kind of spending programmes preferred by particular member governments reflect the importance of individual domestic constituencies, whose interests might be promoted or threatened by particular kinds of EC expenditure. The clearest example of this is agriculture. Community programmes on occasion may usefully diminish pressures on national exchequers. Alternatively the establishment of a Community responsibility for particular kinds of financial programmes may undermine the scope for national governments to devise their own systems of benefits and costs for their domestic clients. The analysis of the way in which decisions are taken and implemented in the EC on financial issues clearly points to major weaknesses in the institutional process. Decisions are taken piecemeal and priorities are not clearly articulated, and indeed all the tendencies within the process are to disaggregate rather than to permit coherence. Stumbling efforts are in hand to correct these weaknesses, but without yet qualitatively altering the nature of the process. One of the factors revealed in the EC context is the absence of a clear distinction between the 'advocates' of public expenditure and the 'guardians' of the public purse. The Commission plays both roles, as do member governments both individually and collectively. This produces a confusion of motives on the part of the actors involved in the budgetary process, which in turn complicates negotiations. It also makes it extremely difficult to determine where and how the process might be rendered more coherent.[6]

Obviously a continuing subject of major interest is the distribution of powers among the EC institutions engaged in the budgetary process. Already this has altered to a remarkable degree since 1970 by comparison with the 1960s, when the mere suggestion of an active role for the European Parliament could serve to provoke a political crisis in the EC. The walls have now been breached in that a series of skirmishes between the Council and the Parliament have made it clear that the Parliament's budgetary powers are not purely symbolic. Appropriations have been altered as a direct consequence of parliamentary interference. However, it would be premature to deduce that Parliament is now poised to secure a firm control over the EC budget as a whole. On the contrary it has not yet succeeded in making any impression at all on the expenditure on price support for agriculture, except marginally to call for more careful control of past expenditure. In the absence of a successful attack on the programme that still consumes around two-thirds of the resources of the EC budget the Parliament still falls well short of acquiring 'the power of the purse'. Indeed it would perhaps be misguided to expect all that much of the European Parliament, when the evidence from national parliaments within the member states illustrates how limited is their control over national budgets. However, it can alternatively be argued that this analogy may not be the most relevant. The record of Congress in the United States in maintaining an active influence on federal appropriations may be a more useful guide to the kind of role that the European Parliament may develop in the future.

More broadly there is a far more fundamental issue about the politics of the EC, namely, how far pecuniary interests are central to the basic political compact struck among the participating governments. In the early years of the EC the political objectives of the founding fathers lay elsewhere with only the occasional intrusion of more immediately materialistic gains that might carry political weight, though it was always asserted that material benefit would reinforce political commitment.[7] Some would still argue that this remains so. However, the political benefits of the EC derive, according to some, primarily from the various advantages of broad political solidarity of a non-quantifiable kind. There are, however, two major difficulties in this dissociation between high politics and economic welfare.[8] The first is that the basic compact among six member states has now been stretched to take in new members, the perceptions of which may be less firmly tied to the old Community orthodoxy. If, as happens to be the case, new members include countries with structurally different economies, the initial compact may carry rather less appeal than it did to the six founder members. Thus their interests may turn to focus on different elements including the effects of EC finances, as has already occurred in the case of the UK.

The second problem is that the financial gains and losses that apparently result from the working of the EC budget are in some sense measurable, even though objectivity of assessment is manifestly absent. By contrast, broader political gains are less easily identified or presented to domestic opinion. In the absence of clear criteria for their evaluation, the narrower financial issues assume a greater importance. Thus when the financial flows become perversely related to the economic capacities of individual member states, this breeds a sense of grievance and injustice, which in turn affects the degree of political commitment expressed by governments and their constituents towards the Community enterprise. The ensuing controversy then in turn affects the wider climate of Community negotiations and diminishes the political solidarity on which the broader compact rests. So while it can be argued that member states may have to suffer financial losses to reap broader political rewards, it does not follow that the losses should necessarily be borne heavily by those who did not help to shape the initial compact. On the contrary it may be that the stability of the system depends on a major recasting of the contract among the member states, with the original signatories adjusting the balance of the *acquis communautaire* to meet the distinctive interests of more recent entrants.

All these issues assume even more importance as the EC enter yet another phase of enlargement. The new member states, as they accede to the EC, will bring an extra set of claims on the limited financial resources of the EC. Simply to extend current expenditure programmes to Greece, Portugal and Spain will require a substantial expansion of the budget.[9] However, the new members will also present new claims on the Community purse, not least for agricultural policy in respect of Mediterranean products. The effect of their entitlement to a share of other categories of expenditure will also alter their availability to existing member states, unless the revenue base is expanded significantly. This can be achieved only by digging deeper into the pockets of the member governments themselves, whether directly or indirectly. Moreover, since all three of the prospective new members would rank as 'less prosperous', their accession to the EC is likely to emphasise the argument about whether the EC budget should be a vehicle for transfers of resources of a redistributive character. The record of the last few years stands witness to the difficulties of the EC in coming to terms with problems of this kind. Conversely it can be argued that the issues have become so pressing that further enlargement could act as a catalyst for a more fundamental reappraisal of the role and objectives of EC financing.

For the student of the EC the budgetary process offers fertile ground for further research. Many of the arguments in both economics and political science about its evolution rest on very scanty and disputed evidence. There is considerable room for clarification of the technical issues that have impeded the establishment of clear criteria for evaluating Community finances. Far more detailed studies need to be made of particular areas of Community expenditure both from the budget and from other financial instruments. Only then will it be possible to judge more systematically the economic impact of the Communities and the political interactions that link the national and Community levels. Yet if the basic hypothesis of this study is correct more rigorous analysis of the financial activities of the EC would yield important insights into the nature of the Community experiment.

Notes: Chapter 5

1   This is one of the chief arguments made by the Cambridge Economic Policy Group, notably in articles in the *Cambridge Economic Review*.
2   The MacDougall Report, Vol. I, esp. pp. 43–72.

3   See Loukas Tsoukalis, *The Politics and Economics of European Monetary Integration*, London, Allen & Unwin, 1977, pp. 90–8.

4   This has been a central issue in debate within the Economic Policy Committee of the EC, widely reported in the press. See, for example, *The Economist*, 11 and 18 November 1978.

5   This is reflected in the German attachment in a federal context to the principle of *Mitspracherrecht* as a process of active consultation between the providers and recipients of revenue.

6   See A. Wildavsky, *Budgeting: A Comparative Theory of Budgetary Processes*, Boston, Mass., Little, Brown, 1975, pp. 7–9 and 187–99. It must be noted, however, that the 'deviant' cases discussed by Wildavsky are where one or both roles are absent or latent, rather than the EC case where both positions are articulated but often by the same actors.

7   This argument figured prominently in the 1975 referendum campaign in the UK following renegotiation, when much of the pro-European literature was designed to demonstrate the cash benefits of EC membership, while the anti-marketeers divided their case. See David Butler and Uwe Kitzinger, *The 1975 Referendum*, London, Macmillan, 1975, pp. 91 and 160–189.

8   See, for example, L. N. Lindberg and S. A. Scheingold, *Europe's Would-Be Polity*, Englewood Cliffs, NJ, Prentice-Hall, 1970, pp. 1–63; and R. Pryce, *The Politics of the European Community*, London, Butterworth, 1973, pp. 9–12.

9   See 'Enlargement of the Community: economic and sectoral aspects', *Bull. EC*, Supplement 3/78, pp. 37–45.

Select Bibliography

Literature specifically on the finances of the European Communities is very sparse. The subject has only recently assumed an important place in political and economic analyses of West European integration. Anyone who wishes to explore in detail particular aspects of Community finances must, therefore, rely heavily on the various official documents published by Community institutions. The more important of these are included in this bibliography.

Commission of the European Communities, Brussels:

Annual reports on the European Agricultural Guidance and Guarantee Fund, on the European Development Fund, on the European Regional Development Fund, and on the European Social Fund.

Preliminary drafts and final versions of the budget for each financial year.

Global Appraisal of the Budgetary Problems of the European Community, produced each spring.

'Financing the common agricultural policy – independent revenue for the Community – wider powers for the European Parliament', *Bull. EEC*, Supplement 5, 1965.

Proposal of the Commission for the Creation of Own Resources for the Communities, Doc. 99, 1969/70-COM(69)700, 16 July 1969.

The Unacceptable Situation and the Corrective Mechanism, Doc. COM(75)40.

'Financing the Community budget: the Way Ahead', *Bull. EC*, Supplement 8, 1978.

These documents make interesting reading and touch on the major issues surrounding the budget, though often in latter years only obliquely.

D. Coombes (ed.), *The Power of the Purse*, London, Allen & Unwin, 1975: a useful set of essays on the budgetary activities of national parliaments in Western Europe.

D. Coombes, *The Future of the European Parliament*, London, Policy Studies Institute, 1979: one of the best recent studies on the Parliament and its evolution past and potential, with many references to its budgetary role.

D. Coombes and I. Wiebecke, *The Power of the Purse in the European Communities*, London, RIIA/PEP, 1972: a survey of developments leading up to the 1970 Treaty and of its implications.

Court of Auditors of the European Communities, Luxembourg, annual reports.

J. E. Danziger, *Making Budgets*, London, Sage Library of Social Research, 1978: an interesting attempt to explore different theoretical approaches to budgets and to apply them to British local government.

G. Denton, 'Reflections on fiscal federalism', *Journal of Common Market*

Studies, vol. XVI, no. 4, June 1978: a commentary on the MacDougall Report from a federalist perspective.

G. Denton, J. Dodsworth, T. Josling and M. Miller, *The Economics of Renegotiation*, London, Federal Trust, May 1975: largely focuses on the budgetary issue and examines the calculations made in Britain on budget contributions both before and after accession.

J. Dodsworth, 'European Community financing: an analysis of the Dublin Amendment', *Journal of Common Market Studies*, vol. XIII, no. 3, December 1975: a somewhat sceptical commentary on the Financial Mechanism.

M. R. Emerson and T. W. K. Scott, 'The Financial Mechanism in the budget of the European Community: the hard core of the British "renegotiations" of 1974–75', *Common Market Law Review*, 14, 1977: a detailed account by two Commission officials written in an optimistic vein.

European Investment Bank, Luxembourg, annual reports; and *20 Years 1958–1978*, 1978: the official history of the EIB, useful as a source in the absence of any more scholarly study, but very bland.

European Parliament, Luxembourg, *The case for a European Audit Office*, 1973: a useful collection of documents and commentary on the budget and in particular on the need to improve financial control, with reference also to national practices; *Purse-strings of Europe; The European Parliament and the Community Budget*, London Office, 1979: pamphlet setting out simply the Parliament's role; reports of the Budget Committee (esp. the Dankert Report, October 1979); *Les ressources propres aux Communautés européennes et les pouvoirs budgétaires du Parlement européen*, 1970: a valuable compilation of documents and comments.

J. Forman, 'The conciliation procedure', *Common Market Law Review*, 16, 1979: argues that the Parliament's powers are gradually extending into the legislative field as a result of its budgetary responsibilities.

W. Godley, 'Policies of the EEC', *Cambridge Economic Review*, April 1979: a highly critical interpretation of the budgetary consequences for member states, especially the UK, of Community finances; important because of the sophistication of the economic analysis by comparison with many other studies, but to be treated with caution because the information used is not comprehensive.

M. Hodges (ed.), *Economic Divergence in the European Community* (provisional title), London, Allen & Unwin, forthcoming: a collection of essays on the changing economic context of the Communities; especially useful for the various discussions of the scope for and relevance of resource transfers, and for the chapter on the budget itself.

E.-S. Kirschen with H. S. Bloch and W. B. Bassett, *Financial Integration in Western Europe*, New York, Columbia University Press, 1969: now very much out of date, but interesting as a commentary on the financial evolution of the EC during the 1960s, before they began to develop a budget.

The MacDougall Report, *The Role of Public Finance in the European Communities*, Brussels, Commission of the EC, 1977: report of a committee of economists on the possible evolution of the EC budget from federalist premises. Volume I contains the report itself, and Volume II a series of more academic articles on budgeting in existing federations. For a summary see also Sir D. MacDougall, 'The future of the EEC Budget', *New Europe*, Autumn 1977, and for a variety of economists' comments see *The MacDougall Report: The Role of Inter-regional Flows of Public Finance in the European Community*, Brussels, Groupes d'Etudes Politiques Européennes, Report No. 3, 1979.

W. A. Oates, *The Political Economy of Fiscal Federalism*, Lexington, Mass., Lexington Books, 1977: a valuable collection of essays on both theory and practice in the analysis of public finance in federations and in multi-level government; includes a chapter on the EC.

J. Rideau, *La France et les Communautés Européennes*, Paris, Librairie Générale de Droit et de Jurisprudence, 1975: includes chapters by P. Manin and P. Amselek on French attitudes to the EC budget.

C. T. Sandford, *The Economics of Public Finance*, Oxford, Pergamon, 1977. A straightforward textbook on the general analysis of public finance.

Second Report of the Select Committee on the European Communities of the House of Lords, Session 1979/80: fairly detailed account of the 1979 budget dispute, the nature of the British grievance and of the early work of the Court of Auditors.

M. Shaw, *The European Parliament and the Community Budget*, Luxembourg, European Conservative Group, 1978: pamphlet by a British member of the European Parliament.

Spaak Committee report, *Report from the Delegation Heads to the Ministers of Foreign Affairs*, Brussels, 1956: worth consulting for the initial views of the founder member states of the EC on Community financing.

D. Strasser, *The Finances of Europe*, New York, Praeger, 1977 (French edn, Paris, Presses Universitaires de France, 1975): a comprehensive account of the development of EC finances and the rules and practices that govern them, but without any general analysis of the underlying political and economic issues.

R. Talbot, 'The European Community's Regional Fund', *Progress in Planning*, vol. 8, no. 3, 1977: detailed account of the creation of the RDF which touches on many of the political disagreements among member states on EC finances.

C. Tugendhat, 'Problems of Community budgeting', *The World Today*, August 1977: essay by the Commissioner for the Budget on some of the issues.

H. Wallace, W. Wallace and C. Webb, *Policy-Making in the European Communities*, Chichester, Wiley, 1977: case studies which touch on Community borrowing, Euratom finance, the Regional Development Fund and VAT harmonisation.

A. Wildavsky, *Budgeting: A Comparative Theory of Budgetary Processes*, Boston, Mass., Little, Brown, 1975: an excellent volume that attempts to develop a comprehensive theoretical framework for the study of budgets, and illustrates it with empirical material from different countries.

Postscript

In the months since this study was completed the budget has dominated discussions within the European Community. It has provided the issue for the newly elected European Parliament to test its powers under the Treaties. It continues as a source of friction between the United Kingdom and the other member states. Increasingly the Common Agricultural Policy has been criticised by the European Parliament and in many member states for the large share of finite budgetary resources that it consumes.

In December 1979 the European Parliament rejected the Community Budget for 1980 by an overwhelming majority, after an unsuccessful request to the Council of Ministers for a symbolic decrease in expenditure on agricultural surpluses, the first time that it had tested the Treaties on its right to modify compulsory expenditure. The Commission and the Council were left to draft a revised budget that would accommodate both the farmers' claims and the Parliament's criticisms.

Commission figures produced in September 1979 demonstrated that in 1980 the United Kingdom would become the largest net provider of Community revenue. The British have argued for an equitable solution. Both the Commission and the British Government have proposed solutions to operate in 1980 ranging from an unconstrained Financial Mechanism to special programmes of Community expenditure in the UK. Two more long term solutions have been mooted: first, a mechanism that would permanently guarantee a defined share of budget receipts for the UK; and secondly, restructuring of the Community budget to reduce the share taken by agriculture and increase other programmes. The first requires a refinement of the criteria for Community expenditure. The second presupposes a willingness by the agriculturally oriented member states either to hold down farm incomes or to accept that they cannot remain wholly dependent on the Community exchequer. Successive European Councils in 1979 and 1980 concentrated on these issues. Their permanent resolution depends on substantial adjustments in both national policies and Community practice.

The threat of the Community running out of money to pay its bills has slightly receded. Current estimates suggest that the 1 per cent ceiling on Value Added Tax may not be reached until 1982. But enlargement and other potential claims on resources will put further pressure on Community funds. The Commission has proposed to devolve more of the costs on to the agricultural sector itself through an increased co-responsibility levy, and to introduce a

Community levy on oil imports and perhaps other forms of energy tax. These would both diversify and extend the resource base of the Community.

The budget has thus emerged as a consuming focus of political attention. It is not the only key issue, but it symbolises the central preoccupations of member governments about the nature of the Community compact and poses awkward dilemmas about the distribution of authority between Community institutions and the member states.

Index

accountability 33, 103–5
acquis communautaire 25, 58, 110
additionality 47–9
appropriations 34, 67, 79, 80–1, 85–6,
88, 109; cash limits 85–6; estimating
problems 46, 66–7; payments and
commitments 68; provisional twelfths
88; *see also* compulsory and non-
compulsory expenditure; maximum
rate
Audit Board 101, 102

Belgium 28–9, 51, 53, 57, 61, 63, 65,
71 n.11, 85

Commission 34, 38, 55, 56, 58, 59, 60,
64, 65–6, 67, 70, 76, 78–83, 84, 88,
89, 91, 94–8, 100, 101, 103, 109;
borrowing 44–5; Budget
Commissioner 78, 82, 98; 1965
proposal 53–4
Committee of Permanent Representatives
83
Common Agricultural Policy 18, 20, 21,
25, 26–7, 41, 42, 44–5, 46, 56–7,
62, 67, 69, 74, 82, 91, 94–5, 97,
107, 109; butter exports to USSR 95;
see also EAGGF
Common External Tariff 57
common market 22, 46, 106
compulsory and non-compulsory
expenditure 67, 76, 77, 84, 87–9
concertation procedure 76, 77, 84
conciliation procedure 87
concurrent studies 36 n.16, 107; *see also*
convergence; EMS
conditionality 47, 48–9, 108
convergence 19, 21–5, 59, 62, 90, 107;
prosperous and less prosperous
countries 22, 23, 24, 50, 88, 107,
111; *see also* divergence
co-ordination of funds 75–6, 78–9, 82,
84–6, 89–90, 97–8, 101, 108–9

Council of Ministers 18, 54, 66, 67–8,
76, 77–8, 79, 80–1, 82, 83–6, 87–9,
90, 91, 97, 99, 102, 103–4, 109;
Agriculture – 85; Budget – 78, 83–6,
89; General – 86; Joint Council of
Foreign and Finance Ministers 78,
86; majority voting 67–8, 83–4, 89,
94–5; Presidency 84
Court of Auditors 77, 80–1, 99, 101–5
Court of Justice 33, 72 n.24, 80–1, 88
customs duties 53–4, 55–7, 60, 61, 62,
83
customs union 21, 55

Danziger, James E. 14
Denmark 28–9, 32, 51, 57, 58, 61, 63,
65, 85, 88
divergence 22, 24, 25, 107; *see also*
convergence

energy policy 25, 41, 43, 44–5, 52–3,
65
enlargement: 1973 23, 25, 58–60, 65,
110–11; Mediterranean 23, 25, 59,
60–2
equity 28–32, 47, 50, 53, 54, 59–60,
62, 83, 91, 110
Euratom 44–5, 52, 53, 55, 65, 76, 85,
101; Supply Agency 44–5, 52
European Agricultural Guidance and
Guarantee Fund (EAGGF) 40, 41,
43, 44–5, 51, 53, 54, 56, 87,
105 n.7; *see also* CAP
European Coal and Steel Community
(ECSC) 20, 39, 41, 44–5, 52, 54–5,
65, 76, 85, 101; High Authority 20,
52, 76
European Council 65, 84; Dublin 1975
59
European Currency Unit (ECU) 68
European Development Fund 39, 41,
44–5, 52, 53, 57, 85; *see also*
overseas aid
European Economic Community (EEC)
52, 53, 55, 76, 101